Lecture Notes
in Business Information Processing   454

Series Editors

Wil van der Aalst🅓
  RWTH Aachen University, Aachen, Germany

John Mylopoulos🅓
  University of Trento, Trento, Italy

Sudha Ram🅓
  University of Arizona, Tucson, AZ, USA

Michael Rosemann🅓
  Queensland University of Technology, Brisbane, QLD, Australia

Clemens Szyperski
  Microsoft Research, Redmond, WA, USA

More information about this series at https://link.springer.com/bookseries/7911

Danielle Costa Morais · Liping Fang (Eds.)

Group Decision and Negotiation

Methodological and Practical Issues

22nd International Conference
on Group Decision and Negotiation, GDN 2022
Virtual Event, June 12–16, 2022
Proceedings

Editors
Danielle Costa Morais (iD)
Universidade Federal de Pernambuco
Recife, Brazil

Liping Fang (iD)
Ryerson University
Toronto, ON, Canada

ISSN 1865-1348   ISSN 1865-1356 (electronic)
Lecture Notes in Business Information Processing
ISBN 978-3-031-07995-5   ISBN 978-3-031-07996-2 (eBook)
https://doi.org/10.1007/978-3-031-07996-2

This Springer imprint is published by the registered company Springer Nature Switzerland AG
The registered company address is: Gewerbestrasse 11, 6330 Cham, Switzerland

Preface

The series of International Conferences on Group Decision and Negotiation (GDN) have furnished a stimulating venue for the dissemination of the latest research on the theory and practice of group decision and negotiation. Conference participants exchange, discuss, and critique their latest ideas in the field of GDN. The GDN conference has been held annually since 2000 with two exceptions: 2000 (Glasgow, UK), 2001 (La Rochelle, France), 2002 (Perth, Australia), 2003 (Istanbul, Turkey), 2004 (Banff, Canada), 2005 (Vienna, Austria), 2006 (Karlsruhe, Germany), 2007 (Mont Tremblant, Canada), 2008 (Coimbra, Portugal), 2009 (Toronto, Canada), 2010 (Delft, The Netherlands), 2011 (cancelled), 2012 (Recife, Brazil), 2013 (Stockholm, Sweden), 2014 (Toulouse, France), 2015 (Warsaw, Poland), 2016 (Bellingham, USA), 2017 (Stuttgart, Germany), 2018 (Nanjing, China), 2019 (Loughborough, UK), 2020 (Toronto, Canada—conference cancelled because of the COVID-19 pandemic and proceedings published), and 2021 (virtual conference hosted from Toronto, Canada).

The 22nd International Conference on Group Decision and Negotiation (GDN 2022) took place virtually (online) during June 12–16, 2022. A total of 68 submissions were received, encompassing nine main streams related to the field of GDN. After a careful and thorough review process involving the work of many reviewers, nine papers from those submissions were selected for publication in this volume entitled "Group Decision and Negotiation: Methodological and Practical Issues."

The nine papers published in this volume are organized into three sections, representing the variety of methodological and practical issues discussed at GDN 2022:

- The first section on "Preference Modeling for Group Decision and Negotiation" consists of four papers. Karpov et al. extend the Hotelling-Downs model in three directions: analyzing competition under a variety of voting rules, considering the circular domain and not just the linear, and extending the results to the winner-take-all setting. Correia et al. analyze the performance of a preference elicitation protocol for multi-issue negotiations based on the Flexible and Interactive Tradeoff (FITradeoff) method through simulation experiments. Suzuki and Horita present two approaches to explore how to reach a global Pareto optimal allocation of robots (indivisible objects) based on the local preference profile. Roselli and de Almeida describe modulation provided by the FITradeoff method from behavioral studies performed using neuroscience tools conducted in two ways: modulation in the preference modeling process and modulation in the FITradeoff Decision Support System (DSS).
- The second section of this volume is composed of three papers related to "Conflict Resolution." Suzuki and Leoneti propose a way to solve the general equilibrium selection problem by means of a meta-level structure, in which each player is assumed to select an equilibrium as their principle of behavior and that such a selection itself makes another game. Philpot et al. use the Graph Model for Conflict Resolution (GMCR) to investigate conflict over expanding aggregate mining at the Alliston

Aquifer in the Township of Tiny, Ontario, shedding light on a growing class of multi-jurisdictional resources conflict. Sharma et al. discuss the use of the concepts of path dependency and leverage points to better understand the complex Cauvery River dispute and recommend the most effective leverage points, including changing the rules of the system as well as the paradigm out of which the system arises.

• The last section on "Collaborative Decision Making Processes" contains two studies of different strategic decision processes. Tanaka et al. apply sentiment analysis to extract the positive, negative, and neutral comments posted on news sites to visualize the social atmosphere by calculating the approval rating of the people's comments. The methodology is demonstrated by analyzing postings regarding COVID-19. Gerchak extends the buy-sell procedure of dissolving two-party partnerships to dissolve three-party partnerships.

We would like to take this opportunity to express our sincere appreciation to many people for their work in organizing GDN 2022 and preparing this volume. Particularly, special thanks go to the Honorary Chair, Rudolf Vetschera, and the other General and Program Chairs, Fuad Aleskerov, Adiel Teixeira de Almeida, and Alexander Lepskiy, for their contributions in organizing GDN 2022 and to the Group Decision and Negotiation (GDN) Section at the Institute for Operations Research and the Management Sciences (INFORMS) in general. We are also thankful to all of the other Stream Organizers: Keith W. Hipel and D. Marc Kilgour (Conflict Resolution); Tomasz Wachowicz (Preference Modeling for Group Decision and Negotiation); Bilyana Martinovski (Emotion in Group Decision and Negotiation); Mareike Schoop, Muhammed-Fatih Kaya, and Rudolf Vestchera (Negotiation Support Systems and Studies (NS3)); Pascale Zaraté (Collaborative Decision Making); Haiyan Xu, Shawei He, and Shinan Zhao (Risk Evaluation and Negotiation Strategies); Zhen Zhang, Yucheng Dong, Francisco Chiclana, and Enrique Herrera-Viedma (Intelligent Group Decision Making and Consensus Process); and Fuad Aleskerov (Social Choice/Manipulability of Aggregation Procedures).

We would like to sincerely thank the reviewers for their informative and prompt reviews of papers: João Clímaco, Ana Paula Costa, Miłosz Kadziński, Ginger Ke, Hannu Nurmi, Leandro C. Rego, Ewa Roszkowska, Maisa Silva, Honorata Sosnowska, Rudolf Vetschera, Junjie Wang, Haiyan Xu, Yi Xiao, and Shinan Zhao.

We are also grateful to the team at Springer for their excellent work.

April 2022

Danielle Costa Morais
Liping Fang

Organization

Honorary Chair

Rudolf Vetschera   University of Vienna, Austria

General Chairs

Fuad Aleskerov	National Research University Higher School of Economics, Russia
Adiel Teixeira de Almeida	Federal University of Pernambuco, Brazil
Liping Fang	Ryerson University, Canada

Program Chairs

Danielle Costa Morais	Federal University of Pernambuco, Brazil
Alexander Lepskiy	National Research University Higher School of Economics, Russia

Program Committee

Melvin F. Shakun	New York University, USA
Adiel Teixeira de Almeida	Federal University of Pernambuco, Brazil
Amer Obeidi	University of Waterloo, Canada
Bilyana Martinovski	Stockholm University, Sweden
Bo Yu	Dalhousie University, Canada
Bogumił Kamiński	Warsaw School of Economics, Poland
Danielle Costa Morais	Federal University of Pernambuco, Brazil
Ewa Roszkowska	University of Białystok, Poland
Fran Ackermann	Curtin Business School, Australia
Fuad Aleskerov	National Research University Higher School of Economics, Russia
Gert-Jan de Vreede	University of South Florida, USA
Ginger Ke	Memorial University of Newfoundland, Canada
Haiyan Xu	Nanjing University of Aeronautics and Astronautics, China
Hannu Nurmi	University of Turku, Finland
João Clímaco	University of Coimbra, Portugal
John Zeleznikow	Victoria University, Australia
José Maria Moreno-Jiménez	Zaragoza University, Spain

Keith Hipel	University of Waterloo, Canada
Kevin Li	University of Windsor, Canada
Liping Fang	Ryerson University, Canada
Love Ekenberg	Stockholm University, Sweden
Luis Dias	University of Coimbra, Portugal
Maisa Mendonça	Federal University of Pernambuco, Brazil
Marc Kilgour	Wilfrid Laurier University, Canada
Mareike Schoop	University of Hohenheim, Germany
Masahide Horita	University of Tokyo, Japan
Pascale Zarate	Université Toulouse 1 Capitole, France
Przemyslaw Szufel	Warsaw School of Economics, Poland
Raimo Hamalainen	Aalto University, Finland
Rudolf Vetschera	University of Vienna, Austria
Rustam Vahidov	Concordia University, Canada
Sabine Koeszegi	Vienna University of Technology, Austria
ShiKui Wu	Lakehead University, Canada
Tomasz Szapiro	Warsaw School of Economics, Poland
Tomasz Wachowicz	University of Economics in Katowice, Poland
Tung Bui	University of Hawai'i, USA
Yufei Yuan	McMaster University, Canada
Jing Ma	Xi'an Jiaotong University, China

Contents

Preference Modeling for Group Decision and Negotiation

Hotelling-Downs Equilibria: Moving Beyond Plurality Variants

Alexander Karpov[1,2]($\boxtimes$) (iD), Omer Lev[3] (iD), and Svetlana Obraztsova[4]

[1] HSE University, Moscow, Russia
akarpov@hse.ru
[2] Institute of Control Sciences, Russian Academy of Sciences, Moscow, Russia
[3] Ben-Gurion University, Beersheba, Israel
[4] Nanyang Technological University, Singapore, Singapore

Abstract. Hotelling-Downs model is a classic model of political competition and strategizing candidates, almost always analyzed under plurality. Our paper presents a three-pronged development of the Hotelling-Downs model. First, we analyze competition under a variety of voting rules. Second, we consider not only a linear city model, but also a circular city model. Third, unlike most Hotelling-Downs papers, we solve the model under the winner-takes-all assumption, which saves many equilibria, and is more relevant to voting settings. In the case of three and four candidates we have found a measure of the set of equilibria.

Keywords: Spatial voting · Scoring methods · Circular city

1 Introduction

Since in many cases voters' behavior patterns are well known, we would expect candidates to react to each other's attempts to garner more support, until settling in a Nash equilibrium, in which none of them can profit by changing their location. This, of course, is not unique to political settings – candidates, being the set of options voters choose from, can be strategic in various settings. Restaurants try to calibrate their menus to the public's taste in their area, sometimes resulting in a sudden multiplication of a particular restaurant type when it becomes popular. A similar effect can be observed in many mass-produced items (e.g., clothing styles), in pricing of similar items, or in musical styles in major competitions (e.g. the Eurovision).

In the political domain, Downs [13] established the spatial model in which voters and candidates are all located in a metric space (with voters preference orders determined by their distance from each candidate), and used Hotelling's results [19] on facility location in metric spaces to show social choice results. Since that ground-breaking work much work has been done exploring the voting

Alexander Karpov was partially supported by the Basic Research Program of the HSE University. Omer Lev acknowledges support of ISF grants #1965/20 and #3152/20.

D. C. Morais and L. Fang (Eds.): GDN 2022, LNBIP 454, pp. 3–16, 2022.
https://doi.org/10.1007/978-3-031-07996-2_1

spatial model in general, and the Hotelling-Downs model in particular. Indeed, a lot of work specifically on candidate manipulation has been done extensively in this model. Despite this, we should stress that in many cases, candidates are located in particular location on the ideological spectrum due to other, exogenous concerns. Much of the existing work in this domain assumes the voting rules are plurality or variants of it, and usually assumes candidates have a utility they wish to maximize, rather than a winner-take-all approach, more common in the computational social choice framework.

In this work we expand results regarding the winner-take-all approach, extending the scope of strategic candidacy in Hotelling-Downs models beyond the limited voting rules discussed so far, to further our understanding of other voting rules, and in particular, to understand how scoring rule parameters affect the strategic space available to candidates. We also show the results regarding a circular model (which has been analyzed in economic Hotelling models, though not in political ones) – circular distribution of voters can be related with preferences over regional (geographical) development, such as a choice of airport location outside a compact city, or related with preferences over calendar events, such as holidays, etc.

In winner-takes-all approach multiplicity of equilibria is a common feature. In many cases candidates have no incentives to deviate from their sincere position. Moreover, in order to understand how likely are equilibria states to emerge as sincere positions points, we pioneer a measure probability – what is the likelihood that randomly placed candidates will be in a state of equilibrium. This helps to understand, for different voting rules (and equilibrium states thereof), how common is an equilibrium state – are there many of them, or only very few? Since candidates might be located in a particular area of the ideological spectrum due to other issues and may be constrained in leaving them (e.g., a political party usually has some spectrum of opinions within it, but one cannot usually extend this spectrum endlessly), seeing how common equilibria are helps to understand whether we are likely to reach them in an emergent, bottom up, candidate generation.

2 Related Work

There are many surveys of Hotelling-Downs models [17,21], the most recent large-scale one is [29]. All attest to how common this model is in political science use when analyzing viewpoints and politics. We shall only mention results which are relevant to (or contrast with) our approach.

There are three main types of candidates objectives: win objective, rank objective, share/score/vote/support objective. "Economic" versions of the model with candidates wishing to maximize support/votes became popular as continuity of utility function allowed for more straightforward analysis. Surely in political competition discontinuity of payoff is a critical feature. The difference of 1% votes is insensible if it does not change the winner and it is crucial if it leads to a new winner.

If there is no profitable deviation in case of rank and support objectives, then there is no profitable deviation in case of win objective. If we have rank and support objectives Nash equilibrium, then we have also the a win objective Nash equilibrium.

For support objectives the model was solved for the run-off rule [7,18,31], probabilistic run-off (Apostolic Voting) [28], best-worst voting rules (mix of plurality and antiplurality) [9], some classes of scoring rules [8,10] and for k-approval rules [32], though that result is problematic, as k can only be fixed (i.e., veto rule, which is $m-1$-approval, for m being the number of candidates, is not solved in that work, and we show the equilibria for this in this paper).

A more natural winner-takes-all assumption draws much less attention [16]. Chisik and Lemke [11] solved the model for three winning-motivated candidates case, uniform distribution of voters with plurality. In this case there are infinite number of equilibria, and the existence is shown for arbitrary number of candidates.

The "linear city" model was developed by Hotelling and Downs [13,19], but a "circular city" model was initially developed by Salop [26]. This model is a basic industrial organizations model [30]. In non-Hotelling-Downs models, circular preferences have been used. Finite population circular domains, such as cyclic group domains [20], circular domains [27], top-circular domains [1], single-peaked on a circle domains [23] have been shown to be manipulable and lead to dictatorial social choice functions under various conditions. Infinite population models do not have these shortcomings. We believe Peeters et al. [22] were the first who presented a Hotelling-Downs model of political competition on a circle. They claimed that far-left and far-right candidates are more similar to each other than candidates in political center.

While outside the scope of this paper, we note that there is some work on strategic candidates outside the Hotelling-Downs model, beginning with [14,15] discussing strategic candidacy in tournaments, and recently further explored by [6,24] and others. This line of work is mainly concerned with candidates deciding if to run or not (and the equilibria this may bring about). Another approach is viewing addition and removal of candidates as a form of control manipulation, studied by [3] (see summaries in [5,25]).

3   Model

We describe the standard one-dimensional spatial model of voting. The policy space is assumed to be a closed $[0,1]$ interval. Let a finite set $C = \{1, \ldots, m\}$ be the set of candidates. Each candidate chooses a point on $[0,1]$ interval.

The set of voters is characterized by distribution of their ideal points μ : $[0,1] \to \mathbb{R}$ with $\int_0^1 d\mu = 1$ (it is assumed, that μ has support set $[0,1]$ and it has no mass points). Each voter has complete transitive preferences over the set of candidates (linear order): let $\pi(C)$ be the set of all full orders over C. For each point $v \in [0,1]$, the Euclidean distance from v defines a weak order of candidates (the closer the better). Two or more candidates may choose the same point. In this case we assume that all possible permutations of these candidates arise in voters preferences with equal probability. For each preference order $P \in \pi(C)$, we define an indicator function $I^P : [0, 1] \rightarrow [0, 1]$, such that $I^P(v) = 1/k$ if and only if there are k linear orders which are linearizations of the weak order of candidates defined by the Euclidean distance from v. Since we are dealing with a potentially infinite number of voters, a preference profile will be defined as a tuple $\mathcal{P} = (I, \mu)$, where I is a tuple of all indicator functions. Let $\mathcal{M}$ be the set of all preference profiles.

Let Q_α be the quantile of order α for voters' distribution μ, formally, we have $\int_0^{Q_\alpha} d\mu = \alpha$.

We consider the following game. m candidates independently and simultaneously choose their positions on a unit interval or on a unit circle. The winner is determined according to a voting rule. All candidates are motivated only by winning. The candidate's first best is to be a sole winner, the second best is to be a winner in the two-winners set, the third best is to be a winner in three-winners set, etc. All losing outcomes are worse and indistinguishable. We distinguish con-vergent Nash equilibrium (CNE), in which all agent have the same position, and non-convergent Nash equilibrium (NCNE), where some candidates have different position.

A voting rule is a function $F : \mathcal{M} \rightarrow 2^C \backslash \emptyset$ from the set of preference orders with their respective measures, obtained from the distribution of voters, to a set of candidates. As usually in Hotelling-Downs models, we consider equilibria in which voters vote sincerely.

We consider the following voting rules:

Defintion 1. A **Scoring rule** chooses a candidate with the highest sum of scores according to a score vector $s = (s_1, s_2, \ldots, s_{m-1}, 0)$, where for each $i \in \{1, \ldots, m-1\}$ we have $s_i \geq s_{i+1}$. The winner is

$$\operatorname*{argmax}_{x \in C} \int_0^1 \sum_{P \in \pi(C)} I^P(v) Score(x, P) \, dv$$

where function $Score(x, P) = s_i$ if and only if candidate x is on position i at preference order P.

We shall specifically mention these scoring rules:

Plurality. The scoring vector $(1, 0, \ldots, 0)$.
Veto. The scoring vector $(1, 1, \ldots, 1, 0)$.
2-approval. The scoring vector $(1, 1, 0 \ldots, 0, 0)$.
Borda. The scoring vector $(m - 1, m - 2, m - 3, \ldots, 1, 0)$ (i.e., $s_i - s_{i+1} = 1$).

Defintion 2. **Scoring elimination rule** is an iterative rule that is based on a scoring rule. In the elimination rule, each iteration, a single candidate with the lowest scores (ties broken by some tie-breaking rule) is eliminated. The last candidate standing is the winner.

For $m = 3$, the **run-off rule** is a special case of the scoring elimination rule with plurality scores $\alpha_1 = 1$, $s_2 = s_3 = 0$.

Defintion 3. *Kemeny rule chooses a candidate at the top of a preference order with the lowest distance to all other preference orders in a profile according to swap distance metrics*

$$Topcandidate \left(\underset{K \in \pi(C)}{argmin} \sum_{P \in \pi(C)} \int_0^1 I^P(v) Swaps(P, K) \, dv \right),$$

where function $Swaps(P, K)$ is the swap distance between two preference orders (i.e., Kendall-Tau metric).

3.1 CNE. Known Results

CNE with different objectives coincide.

Proposition 1 *(Cox, 1987) [12]. For a given scoring rule s, there exists CNE, in which all candidates choose Q_α, if and only if $c(s, m) \leq \alpha \leq 1 - c(s, m)$, where $c(s, m) = \frac{s_1 - (1/m) \sum_{k=1}^m s_k}{s_1 - s_m}$.*

Since median point satisfies condition from Proposition 1 it is a generalization of the median voter theorem. In particular CNE exists for the veto rule and the Borda rule.

Proposition 2 *(Cox, 1987) [12]. For the Condorcet-consistent rules there is a CNE with all candidates located in the median voter position. There is no other equilibrium.*

3.2 NCNE Known Results

The literature of NCNE with support objectives for scoring rules [8–10,12] consist of many existence/characterization results, in which candidates constitute a special structure, e.g. distribution of candidates with equal distance between neighbouring candidates.

A scoring rule s is convex if $s_1 - s_2 \geq s_2 - s_3 \geq s_{m-1} - s_m$.

Proposition 3 *(Cahan, Slinko 2017) [10]. For a convex scoring rule s such that $s_n > s_{n+1} = \ldots = s_m$, where $1 \leq n \leq m$, there is a NCNE with support objectives if and only if the subrule $s' = (s_1, \ldots, s_n, s_{n+1})$ is the Borda rule and $n + 1 \leq \lfloor m/2 \rfloor$.*

A scoring rule s is symmetric if $s_i - s_{i+1} = s_{m-i} - s_{m-i+1}$, for all $1 \leq \lfloor m/2 \rfloor$.

Proposition 4 *(Cahan, Slinko 2017) [10]. For a symmetric scoring rule s, there is no NCNE with support objectives.*

A scoring rule s is weakly concave if $s_i - s_{i+1} \leq s_{m-i} - s_{m-i+1}$.

Proposition 5 *(Cahan, Slinko 2018) [8]. For a weakly concave scoring rule s has no NCNE with support objectives in which $max(n_1, n_q) \leq \lfloor m/2 \rfloor + 1$, where n_1 and n_q are numbers of candidates in the most left and the most right locations. If? in addition, s satisfies*

$$s_4 + s_{m-3} \geq \frac{1}{m-3}\left(\sum_{i=1}^{m-3} s_i + \sum_{i=4}^{m} s_i\right),$$

then no NCNE with support incentives exist.

These results are too special. We consider win objectives, which allow more equilibria.

4   Voters and Candidates Uniform on a Line

Firstly, we present results a benchmark case of voters uniformly distributed on a line.

Let us consider the scoring rule family with scores $(1, \alpha, 0)$, $\alpha < 1$ ($m = 3$ case). Chisik and Lemke [11] showed that only an extreme candidate wins in equilibrium under the plurality rule in three candidates elections. Lemma 1 generalizes this result for all scoring rules.

For the remainder of this subsection this section we consider a model with three candidates I, II, III located on $[0, 1]$ interval. Candidates I, II, III have positions x, y, z, correspondingly, where $x < y < z$.

Lemma 1. *For a scoring rule, there are no equilibria in which candidate II is a winner.*

Proof. The candidates' scores are shown in Table 1. The Table 1 summarizes partition of voters and scores from six possible linear orders (because of linear space we have only four linear orders with non-zero measure). Using Definition 1 we find a winner.

Table 1. Candidates' scores for Lemma 1

	1 point	α points
Candidate I	Interval $[0, x + \frac{y-x}{2}]$, Length of interval $\frac{y+x}{2}$	Interval $[x + \frac{y-x}{2}, x + \frac{z-x}{2}]$, Length of interval $\frac{z-y}{2}$
Candidate II	Interval $[x + \frac{y-x}{2}, y + \frac{z-y}{2}]$, Length of interval $\frac{z-x}{2}$	Intervals $[0, x + \frac{y-x}{2}] \cup [z - \frac{z-y}{2}, 1]$, Length of intervals $1 - \frac{z-x}{2}$
Candidate III	Interval $[z - \frac{z-y}{2}, 1]$, Length of interval $1 - \frac{z+y}{2}$	Interval $[x + \frac{z-x}{2}, z - \frac{z-y}{2}]$, Length of interval $\frac{y-x}{2}$

If candidate II is a winner then

$$\frac{z-x}{2} + \alpha(1 - \frac{z-x}{2}) > \frac{x+y}{2} + \alpha\frac{z-y}{2},$$

$$\frac{z-x}{2} + \alpha(1 - \frac{z-x}{2}) > 1 - \frac{y+z}{2} + \alpha\frac{y-x}{2}.$$

Summing and simplifying these equations we get

$$z - x > \frac{2}{3} \cdot \frac{1 - 2\alpha}{1 - \alpha}.$$

There should also be no incentive for candidate I to deviate to candidate II's location (y), so candidate III should win if that happens:

$$\frac{1+\alpha}{2} \cdot \frac{z+y}{2} + \frac{\alpha}{2}(1 - \frac{z+y}{2}) < 1 - \frac{z+y}{2},$$

$$\frac{y+z}{4} + \frac{\alpha}{2} < 1 - \frac{z+y}{2},$$

$$\frac{3}{4}(y+z) < 1 - \frac{\alpha}{2}.$$

Similarly, not allowing candidate III to deviate to y results in $-\frac{3}{4}(x+y) < -\frac{1+\alpha}{2}$. Combining these together we get $\frac{1}{2} - \alpha > \frac{3}{4}(z-x)$, which is simplified to $z - x < \frac{2}{3} - \frac{4}{3}\alpha$. Thus, we have

$$\frac{2}{3} - \frac{4}{3}\alpha > z - x > \frac{2}{3} \cdot \frac{1 - 2\alpha}{1 - \alpha}.$$

Simplified, this results in $1 - 2\alpha > \frac{1-2\alpha}{1-\alpha}$ which cannot be true for any $\alpha < 1$.

We now begin to investigate how the α value changes the equilibria states.

Proposition 6. *For scoring rules with $\alpha \geq \frac{1}{2}$, there is no equilibrium with different locations of candidates.*

Proof. If there are more than two different locations for the candidates in equilibrium, candidate II should not want to deviate to x or to z (as we know from Lemma 1 that it is not the winner). Following the same calculation done in Lemma 1 for moving candidate I to y, this results in: $\frac{3}{4}(x+z) < 1 - \frac{\alpha}{2}$ (x has replaced y in the Lemma's expression). Similarly, candidate II cannot move to z and profit, thus (again, following Lemma 1), $-\frac{3}{4}(x+z) < -\frac{1+\alpha}{2}$. Summing these together, we have $0 < \frac{1}{2} - \alpha$ which is not true for $\alpha \geq \frac{1}{2}$.

Thus for $\alpha \geq \frac{1}{2}$, it is profitable for candidate II to deviate to one of the other candidates. But that means that there is a state where two candidates win, and hence the third would benefit from deviating to them. Thus, an equilibrium would contain a single location.

For a scoring rule with $\alpha < \frac{1}{2}$, equilibrium always exists with different locations of candidates.

4.1 Measure of the Set of Equilibria

Where we know there are equilibria, and yet do not have an easy and concise representation of it, we wish to understand how prevalent are equilibria states in the whole domain. In order to do this, we calculate the measure of the set of equilibria. 1 means that almost every distribution (except the set of measure zero) leads to an equilibrium. 0 means that it may exist, but it requires some special configurations of measure zero. Equilibrium is not a random event. Because of it we do not use term probability here.

Proposition 7. *For scoring rules with $\alpha < \frac{1}{2}$, the measure of the set of equilibria equals to $\frac{8}{27} \cdot \frac{(1-2\alpha)^3}{(1-\alpha)^2}$.*

Proof. Using the same values from Table 1, suppose candidate I is a winner:

$$\frac{x+y}{2} + \alpha\frac{z-y}{2} > \frac{z-x}{2} + \alpha(1 - \frac{z-x}{2}),$$

$$\frac{x+y}{2} + \alpha\frac{z-y}{2} > 1 - \frac{y+z}{2} + \alpha\frac{y-x}{2}.$$

Simplifying we have

$$x(1 - \frac{\alpha}{2}) + y(\frac{1}{2} - \frac{\alpha}{2}) + z(\alpha - \frac{1}{2}) > \alpha,$$

$$x(\frac{1}{2} + \frac{\alpha}{2}) + y(1 - \alpha) + z(\frac{1}{2} + \frac{\alpha}{2}) > 1.$$

There is no profitable deviation of candidate II with staying in the same point as the winner if:

$$\frac{1}{2} \cdot (x + \frac{z-x}{2} + \alpha) < \frac{z-x}{2} + 1 - z.$$

Which, simplified, is $z + x < \frac{4}{3} - \frac{2}{3}\alpha$.

Candidate II also should not be able to deviate profitably to $x < y' < z$, thus, it cannot be that:

$$\frac{z-x}{2} + \alpha(1 - \frac{z-x}{2}) > \frac{x+y'}{2} + \alpha\frac{z-y'}{2},$$

$$\frac{z-x}{2} + \alpha(1 - \frac{z-x}{2}) > 1 - \frac{y'+z}{2} + \alpha\frac{y'-x}{2}.$$

Summing those together, we get $z - x > \frac{2}{3}\frac{1-2\alpha}{1-\alpha}$, and since this should not happen, we will wish to maintain $z - x \leq \frac{2}{3} \cdot \frac{1-2\alpha}{1-\alpha}$. Candidate II should also not profit from moving to $y' < x$, thus it cannot be that:

$$\frac{x+y'}{2} + \alpha\frac{z-x}{2} > \frac{z-y'}{2} + \alpha(1 - \frac{z-y'}{2}),$$

$$\frac{x+y'}{2} + \alpha\frac{z-x}{2} > 1 - \frac{x+z}{2} + \alpha\frac{x-y'}{2}.$$

Summing together, and since the value for candidate II will increase the closest it gets to x, we get $x(3 - \frac{3}{2}\alpha) - \frac{3}{2}\alpha z > 1 + \alpha$, so for this not to be possible, $x(3 - \frac{3}{2}\alpha) - \frac{3}{2}\alpha z \leq 1 + \alpha$. Similarly, candidate II cannot become the winner due to moving to $y' > z$, so $z(3 - \frac{3}{2}\alpha) + \frac{3}{2}\alpha x \leq 2 - \alpha$.

A similar process for deviations of candidate III leads to additional constraints. However, the ones that are binding (many constraints are repetitive) are:

$$x(\frac{1}{2} + \frac{\alpha}{2}) + y(1 - \alpha) + z(\frac{1}{2} + \frac{\alpha}{2}) > 1;$$

$$y < z; z + x < \frac{4}{3} - \frac{2}{3}\alpha; z - x > \frac{2}{3} \cdot \frac{1 - 2\alpha}{1 - \alpha};$$

$$x(1.5 - \frac{\alpha}{2}) + z(0.5 + \frac{\alpha}{2}) \leq 1$$

The volume of this area equals to:

$$\int_{(2\alpha-\alpha^2)/(1-\alpha)/3}^{(1-\alpha+\alpha^2)/(1-\alpha)/3} \int_{(2-x(1+\alpha))/(3-\alpha)}^{x+(2-\alpha4)/(1-\alpha)/3} \int_{(2-x(1+\alpha)-z(1+\alpha))/(1-\alpha)/2}^{z} 1\, dy\, dz\, dx$$

$$+ \int_{(1-\alpha+\alpha^2)/(1-\alpha)/3}^{\frac{1}{2}} \int_{(2-x(1+\alpha))/(3-\alpha)}^{(2-x(3-\alpha))/(1+\alpha)} \int_{(2-x(1+\alpha)-z(1+\alpha))/(1-\alpha)/2}^{z} 1\, dy\, dz\, dx$$

$$= \frac{2}{81} \frac{(1 - 2\alpha)^3}{(1 - \alpha)^2}.$$

Because there are six ways to rename candidates and two ways locate winner (near left end, or near right end of the interval) there are 12 equivalent cases. The measure of the set of equilibria equals

$$\frac{8}{27} \cdot \frac{(1 - 2\alpha)^3}{(1 - \alpha)^2}.$$

Proposition 8. *For each scoring elimination rule, the measure of the set of equilibria equals to 0.*

Proof. For the third place candidate a strategy of staying in the same point with a winner guarantees a positive probability of winning the first round. On the second round the winner position guarantees positive probability of winning.

Proposition 9. *For each rule based on the majority relation, the measure of the set of equilibria equals to 0.*

Proof. Median candidate position deviation is profitable.

5 Voters and Candidates Uniform on a Circle

In this section we assume we have the unit circle, that is the point 0 and the point 1 are the same point. Note that thanks to this being a circle we can decide where the point 0 is on it.

5.1 Existence of Equilibrium

When analyzing single-peaked preferences on a circle, one of the important observations is that there is no median voter position. The equidistant distribution of candidates has direct counterpart in Salop circular city model [26]. It is quite straightforward to see that an equidistant distribution of candidates is an equilibrium under plurality rule and a run-off rule. The veto rule adds several equilibria – there is one where all candidates are at the same point, but also for an even m, candidates can be divided equally and located centrosymmetrically in two locations. There are also other equilibria. Furthermore, all rules based on the weighted tournament matrix have equilibria for each location of candidates.

For the plurality rule we completely solve the cases of $m = 3$ and $m = 4$.

5.2 Uniform on a Circle, $m = 3$

Since a circle has no fixed location anchor, let us mark a candidate I's position as 0. A direction towards candidate II is the direction of the axis. The position of candidate II is y and the position of candidate III is z.

Because of centrosymmetry the probability of each ranking equals to probability of reverse ranking $Pr(123) = Pr(321)$, $Pr(132) = Pr(231)$, $Pr(213) = Pr(312)$.

Proposition 10. *For each scoring rule, if* $\alpha = \frac{1}{2}$, *the measure of the set of equilibria equals to 1; if* $\alpha < \frac{1}{2}$ *it is* $\frac{1}{6}$; *and if* $\alpha > \frac{1}{2}$ *it is 0.*

Proof. The candidates' scores are shown in Table 2.

Table 2. Candidates' scores for Proposition 10

	1 point	α points
Candidate I	Intervals $[0, \frac{y}{2}] \cup [1 - \frac{1-z}{2}, 1]$, Length of intervals $\frac{y}{2} + \frac{1-z}{2}$	Intervals $[\frac{y}{2}, \frac{z}{2}] \cup [y + \frac{1-y}{2}, 1 - \frac{1-z}{2}]$, Length of intervals $z - y$
Candidate II	Interval $[\frac{y}{2}, y + \frac{z-y}{2}]$, Length of interval $\frac{z}{2}$	Intervals $[\max(0, z - 1 + \frac{1-z+y}{2}), \frac{y}{2}]$ $\cup [y + \frac{z-y}{2}, y + \frac{1-y}{2}]$ $\cup (\min(1, z + \frac{1-z+y}{2}), 1]$, Length of intervals $1 - z$
Candidate III	Interval $[z - \frac{z-y}{2}, 1 - \frac{1-z}{2}]$, Length of interval $\frac{1-y}{2}$	Intervals $[\frac{z}{2}, y + \frac{z-y}{2}]$ $\cup [1 - \frac{1-z}{2}, \min(z + \frac{1-z+y}{2}, 1)]$ $\cup [0, \max(0, z + \frac{1-z+y}{2} - 1)]$, Length of intervals y

If candidate I is a winner then

$$\frac{y}{2} + \frac{1-z}{2} + \alpha(z - y) > \frac{z}{2} + \alpha(1 - z),$$

$$\frac{y}{2} + \frac{1-z}{2} + \alpha(z - y) > \frac{1-y}{2} + \alpha y$$

Transforming, we have

$$y(\frac{1}{2} - \alpha) + z(-1 + 2\alpha) > -\frac{1}{2} + \alpha,$$

$$y(1 - 2\alpha) + z(-\frac{1}{2} + \alpha) > 0$$

In case of $\alpha = \frac{1}{2}$ all candidates have the same scores in each realization of positions. The probability of equilibrium is equal to 1.

There is no profitable deviation of staying in the same point with the winner if $\frac{1}{4} + \frac{\alpha}{2} < \frac{1}{2}$.

In case of $\alpha > \frac{1}{2}$ there is no equilibrium. Let us consider the case of $\alpha < \frac{1}{2}$. There should be no profitable deviation of candidate II to $x < y' < z$, thus:

$$\frac{z}{2} + \alpha(1 - z) < \frac{y'}{2} + \frac{1 - z}{2} + \alpha(z - y'),$$

$$\frac{z}{2} + \alpha(1 - z) < \frac{1 - y'}{2} + \alpha y'.$$

Summing these up and simplifying, we get $z < \frac{2}{3}$. Similarly, there is no profitable deviation of candidate III to $y < z' < 1$, and similarly summing the inequalities will result in $y \geq \frac{1}{3}$. Thus, the measure of the set of equilibria equals to $\frac{1}{6}$.

Proposition 11. *For a scoring elimination rule with $\alpha \neq \frac{1}{2}$, the measure of the set of equilibria equals to 0.*

Proof. For the third ranked candidate deviating to the same point with a winner guarantees a positive probability of winning the first round. On the second round candidates would have exactly the same scores and probability of winning.

Proposition 12. *For the Kemeny rule, the measure of the set of equilibria equals to 0.*

Proof. Because of symmetry we have a pair of winning rankings with equal scores, and are opposite. The third place candidate has beneficial deviation to stay at the same point with a winner, as there is a positive probability of winning.

5.3   Uniform on a Circle, $m = 4$

In a quarter of cases, candidate I is a winner. Let us mark a candidate I's position as 0. We chose direction of the axis such that the position of candidate III is greater or equal than $\frac{1}{2}$. The position of candidate II is y, the position of candidate III is z ($z \geq \frac{1}{2}$), a position of candidate IV is t.

Proposition 13. *For the plurality rule, the measure of the set of equilibria equals to 0.*

Proof. The scores are:

Candidate I $\frac{y}{2} + \frac{1-t}{2}$ **Candidate III** $\frac{t-y}{2}$

Candidate II $\frac{z}{2}$ **Candidate IV** $\frac{1-z}{2}$

Candidate II obviously beats candidate IV (since $z \geq \frac{1}{2}$). Candidate II can deviate to the point $t - \frac{1}{2}$ which will make candidates I, III have equal score $-\frac{1}{4}$, which is below candidate II's score (since $z > \frac{1}{2}$). Thus, there is always a deviation for candidate II, i.e., there is no equilibrium.

Proposition 14. *For the run-off rule, the measure of the set of equilibria equals to 0.25.*

Proof. As shown in Proposition 13, candidate II can always make it to the top 2. Suppose candidates 1 and 2 are the first round winners, then we can write down the equations that maintain that candidates 3 and 4 score less, and thus we reach the constraints:

$$z \geq \frac{1}{2}; 0 \leq y \leq \frac{1}{2}; z \leq t; z \leq 1 - t + y; t \leq 1; t \leq \frac{1}{2} + y$$

The volume of this area equals to

$$\int_0^{\frac{1}{2}} \int_{\frac{1}{2}}^{\frac{1}{2}+y/2} \int_{\frac{1}{2}}^t 1 \, dz \, dt \, dy + \int_0^{\frac{1}{2}} \int_{\frac{1}{2}+y/2}^{\frac{1}{2}+y} \int_{\frac{1}{2}}^{1-t+y} 1 \, dz \, dt \, dy = \frac{1}{96}.$$

Because there are 24 equivalent cases (possible renamings), the measure equals $\frac{1}{4}$.

The following propositions, for which only proof sketches are provided, hold for all m.

Proposition 15. *For the Borda rule, the measure of the set of equilibria equals to 1.*

Proof. Because of centrosymmetry the probability of each ranking equals to probability of reverse ranking. Scores of all candidates are equal due to the linear reduction in score in the vector. There are no incentives for deviation.

Proposition 16. *For each rule based on the majority relation, the measure of the set of equilibria equals to 1.*

Proof. We have ties for all paired comparisons.

Proposition 17. *For the veto rule, the measure of the set of equilibria equals to 0.*

Proof. Standing between the two consecutive candidates with the smallest distance is always profitable.

Proposition 18. *For the 2-approval, the measure of the set of equilibria equals to 1.*

Proof. Each candidate has two neighbour candidates and exactly one opposite candidate. For each pair (candidate-opposite candidate), there are two points in which distances from these candidates are equal. The distance between these points is 1/2. On each path between candidate and opposite candidate there is exactly one candidate. Thus each candidate receive exactly 1/2 approval votes.

6 Discussion

In this paper we have extended Hotelling-Downs settings in three directions: we extend the results on winner-take-all settings; we expand beyond plurality and its variants to many scoring rules, as well as a few non-scoring rules results; and we explore circular domains, and not just the line. Moreover, we have introduced the measure of the set of equilibria. In some sense it characterizes stability of voting rule.

The Hotelling-Downs model results are of particular interest in today's research map – as there is a growing interest in political parties (e.g., gerrymandering [2] or primaries [4]), the shift of parties' candidates in intra-party competitions may be much more pronounced, as the party can choose its candidates to be anywhere on its spectrum of views. We hope this research will help contribute to this topic as well.

There are plenty of open problems still to explore – such as increasing the number of candidates; expanding to more domains beyond the unit circle; and integrating with party-based models. We also believe our metric can be useful in further domains which can benefit from understanding how likely an emergent state is to "accidentally" be an equilibrium.

References

1. Achuthankutty, G., Roy, S.: Dictatorship on top-circular domains. Theor Decis. **86**, 479–493 (2018)
2. Bachrach, Y., Lev, O., Lewenberg, Y., Zick, Y.: Misrepresentation in district voting. In: Proceedings of the 25th International Joint Conference on Artificial Intelligence (IJCAI), New York City, pp. 81–87, July 2016
3. Bartholdi III, J.J., Tovey, C.A., Trick, M.A.: How hard is it to control an election? Math. Comput. Model. **16**(8–9), 27–40 (1992)
4. Borodin, A., Lev, O., Shah, N., Strangway, T.: Primarily about primaries. In: Proceedings of the 33rd Conference on Artificial Intelligence (AAAI), Honolulu, Hawaii, pp. 1804–1811, January–February 2019
5. Brandt, F., Conitzer, V., Endriss, U., Lang, J., Procaccia, A.D. (eds.): Handbook of Computational Social Choice. Cambridge University Press, Cambridge (2016)
6. Brill, M., Conitzer, V.: Strategic voting and strategic candidacy. In: Proceedings of the 29th AAAI Conference on Artificial Intelligence (AAAI), Austin, Texas, pp. 819–826, January 2015
7. Brusco, S., Dziubiński, M., Roy, J.: The hotellingdowns model with runoff voting. Games Econom. Behav. **74**, 447–469 (2012)
8. Cahan, D., McCabe-Dansted, J., Slinko, A.: Asymmetric equilibria in spatial competition under weakly concave scoring rules. Econ. Lett. **167**, 71–74 (2018)
9. Cahan, D., Slinko, A.: Electoral competition under best-worst voting rules. Soc. Choice Welfare **51**(2), 259–279 (2018). https://doi.org/10.1007/s00355-018-1115-7
10. Cahan, D., Slinko, A.: Nonconvergent electoral equilibria under scoring rules: beyond plurality. J. Public Econ. Theor **19**(2), 445–460 (2018)
11. Chisik, R.A., Lemke, R.J.: When winning is the only thing: pure strategy Nash equilibria in a three-candidate spatial voting model. Soc. Choice Welfare **26**(1), 209–215 (2006)

12. Cox, G.W.: Electoral equilibrium under alternative voting institutions. Am. J. Polit. Sci. **31**(1), 82–108 (1987)
13. Downs, A.: An Economic Theory of Democracy. Harper and Row, New York (1957)
14. Dutta, B., Jackson, M.O., Le Breton, M.: Strategic candidacy and voting procedures. Econometrica **69**(4), 1013–1037 (2001)
15. Dutta, B., Jackson, M.O., Le Breton, M.: Voting by successive elimination and strategic candidacy. J. Econ. Theor. **103**(1), 190–218 (2002)
16. Feldman, M., Fiat, A., Obraztsova, S.: Variations on the hotelling-downs model. In: Proceedings of the 30th Conference on Artificial Intelligence (AAAI), Phoenix, Arizona, pp. 496–501, February 2016
17. Grofman, B.: Downs and two-party convergence. Annu. Rev. Polit. Sci. **7**, 24–46 (2004)
18. Haan, M., Volkerink, B.: A runoff system restores the principle of minimum differentiation. Eur. J. Polit. Econ. **17**, 157–162 (2001)
19. Hotelling, H.: Stability in competition. Econ. J. **39**, 41–57 (1929)
20. Kim, K.H., Roush, F.W.: Special domains and nonmanipulability. Math. Soc. Sci. **1**, 85–92 (1980)
21. Osborne, M.J.: Spatial models of political competition under plurality rule: a survey of some explanations of the number of candidates and the positions they take. Can. J. Econ. **28**(2), 261–301 (1995)
22. Peeters, R., Saran, R., Yüksel, A.M.: Strategic party formation on a circle and durverger's law. Soc. Choice Welfare **47**, 729–759 (2016)
23. Peters, D., Lackner, M.: Preferences single-peaked on a circle. In: Proceedings of the 31st AAAI Conference on Artificial Intelligence (AAAI) (2017)
24. Polukarov, M., Obraztsova, S., Rabinovich, Z., Kruglyi, A., Jennings, N.R.: Convergence to equilibria in strategic candidacy. In: Proceedings of the 24th International Conference on Artificial Intelligence (IJCAI), Buenos Aires, Argentina, pp. 624–630, July 2015
25. Rothe, J. (ed.): Economics and Computation. STBE, Springer, Heidelberg (2016). https://doi.org/10.1007/978-3-662-47904-9
26. Salop, S.C.: Monopolistic competition with outside goods. Bell J. Econ. **10**, 141–156 (1979)
27. Sato, S.: Circular domains. Rev. Econ. Design **14**, 331–342 (2010)
28. Savčić, R., Xefteris, D.: Apostolic voting. Can. J. Econ. **54**, 1400–1417 (2021)
29. Schofield, N.: The Spatial Model of Politics. No. 95 in Routledge Frontiers of Political Economy, Routledge (2008)
30. Tirole, J.: The Theory of Industrial Organization. MIT Press (1988)
31. Xefteris, D.: Mixed equilibria in runoff elections. Games Econom. Behav. **87**, 619–623 (2014)
32. Xefteris, D.: Stability in electoral competition: a case for multiple votes. J. Econ. Theor. **161**, 76–102 (2016)

Eliciting Preferences with Partial Information in Multi-issue Negotiations: An Analysis of the FITradeoff-Based Negotiation Protocol

Lucas Miguel Alencar de Morais Correia[1,2], Eduarda Asfora Frej[1,2(✉)],
Manoel Lucas Sousa Ribeiro[1,2], and Danielle Costa Morais[1]

[1] Universidade Federal de Pernambuco, Av. da Arquitetura - Cidade Universitária,
Recife, PE, Brazil
eafrej@cdsid.org.br
[2] Center for Decision Systems and Information Development–CDSID,
Universidade Federal de Pernambuco, Av. da Arquitetura–Cidade Universitária,
Recife, PE 50.630-970, Brazil

Abstract. This paper aims to analyze the performance of the preferences elicitation protocol for multi-issue negotiations based on the FITradeoff multicriteria method through simulation experiments. The negotiation protocol is conducted based on a dynamic set of packages, considering partial information about negotiators preferences. Efficiency of the agreement package can also be verified through a pareto optimality analysis. Simulation experiments involving negotiators were performed, supported by a web-based Electronic Negotiation System (ENS), through which the whole process is operationalized. The results of the experiments show that the FITradeoff-based negotiation protocol enables the parties to achieve an agreement without spending much cognitive effort in the preferences elicitation process. Moreover, in most cases, an agreement package can be faster due to the convergence mechanism derived from the dynamic set approach.

Keywords: Negotiation support · FITradeof method · Preferences elicitation · Partial information

1 Introduction

Negotiation is a complex process that can involve two or more parties and multiple issues in an effort to reconcile often opposing interests [1]. Thus, Multiple Criteria Decision Aiding (MCDA) techniques can be used in the negotiation analysis to support the negotiation parties in eliciting their preferences [2, 3]. Several support methods and software tools were developed to facilitate the negotiation decision-making processes. More precisely, in methodological terms, several multi-criteria decision methods (MCDM) such as AHP [4], TOPSIS [2], ELECTRE-TRI [5], MARS [6, 7] and UTASTAR [1, 7, 8] are applied to help negotiators in the pre-negotiation phase to build offer scoring systems in negotiations [9]. However, a challenge in the MCDM area is to devise a process of obtaining preferences, where decision makers (DMs) feel comfortable and that does not

D. C. Morais and L. Fang (Eds.): GDN 2022, LNBIP 454, pp. 17–30, 2022.
https://doi.org/10.1007/978-3-031-07996-2_2

require much cognitive effort in the elicitation process, in a way that produces preferences that represent their interests [10].

Negotiation Support Systems (NSS) are projected to support negotiators in complex negotiations, including aiding in preference elicitation, conflict management and resolution, seeking consensus assessment and equilibrium analysis [11]. In addition, such systems measure the scale of concessions and allow viewing the progress of the negotiation, being very useful in the quantitative assessment of the negotiation offers [12]. Several formal decision support models for negotiation problems are operationalized in negotiation support systems or Electronic Negotiation Systems (ENS), used in research and negotiation training contexts, such as Inspire [12], Negoisst [13, 14] or NegoCalc [15] and eNego [7], where decision support on offers (packages) provided by the vast majority of NSS/ENS is based on the model of additive aggregation [16].

Frej et al. [10] recently developed a preferences elicitation protocol with partial information based on the FITradeoff method [18, 19]. This approach was mainly developed motivated by the reason that negotiators often may not be able to provide complete information about scoring systems and/or utility functions when eliciting preferences, and sometimes only incomplete information about model parameters is available [17, 20]. In this sense, a negotiation protocol based on the FITradeoff is presented in [10], considering a dynamic set of packages during the negotiation rounds in order to fasten the convergence for an agreement package. The preferences elicitation is conducted during the offer exchanging process, and negotiators provide as much information as they are willing to [10]. Efficiency of the agreement package is also verified considering a pareto optimality analysis.

In this context, this paper aims to analyze the performance of the FITradeoff-based negotiation protocol developed by [10] through an experiment of electronic negotiations simulations conducted based on the FITradeoff-based electronic negotiation system (ENS). This approach presents a structured negotiation process that requires less time and cognitive effort on the part of negotiators for a mutual agreement to be identified, without using scoring systems or utility functions to evaluate negotiation packages, but through a ranking of packages. Therefore, in order to verify the behavior of negotiators in the pre-negotiation and negotiation conduction phases, as well as the convergence to an agreement between the parties, in this work we analyze a set of data obtained through an experiment of electronic negotiations simulations conducted based on a web-based electronic negotiation system built to operationalize such negotiation protocol.

This paper is organized as follows. Section 2 briefly presents the FITradeoff-based negotiation protocol, which is detailed in [10]. Section 3 presents the analysis of bilateral negotiation simulation with the FITradeoff electronic negotiation system FITradeoff, including details of the representation of preferences. Section 4 describes the main results of the negotiation simulation experiments, and Sect. 5 discusses the results obtained, outlines the basic findings in order to draw conclusions and suggest some lines of future research.

2   FITradeoff-Based Negotiation Protocol

2.1   Preferences Elicitation with FITradeoff

The Flexible Interactive Tradeoff - FITradeoff method [18, 19] was developed in order to model the preferences of a decision maker with partial information about the values of the scaling constants within the scope of Multiattribute Value Theory (MAVT), maintaining the entire axiomatic structure [21] of the traditional tradeoff procedure [16]. Criteria (issues) scaling constants are usually also known as criteria (issues) weights; however, their meaning does not have the sense of the degree of importance of the criteria, since the ranges of the values of the consequences of the criteria must be taken into account for the determination of their values [18].

In this method, the applicability of the traditional tradeoff is improved, because the questions asked to the decision maker in the elicitation process are easier, thus requiring less information to be provided [22]. More precisely, in FITradeoff, the decision maker is asked to compare the consequences by giving strict preference statements [18], whereas, in the traditional tradeoff procedure, the decision maker is obliged to establish indifference relations, since it is cognitively more difficult for the decision maker, leading to a high rate of inconsistencies - 67% [25]. Thus, by incorporating the concept of flexible elicitation in the FITradeoff method, it requires less cognitive effort on the part of the decision maker in the preference elicitation process and, consequently, reduces the possibility of inconsistencies.

In its initial proposal, the FITradeoff method focused on solving only multicriteria decision problems in the problematic of choice. Later, Frej et al. [22], extended the concept of flexible elicitation, making it also suitable for solving multicriteria decision problems in the context of the problematic of ranking. In addition to this, other FITradeoff variants were also developed, such as for the sorting problematic [23] and for the portfolio problematic [24]. Furthermore, the incorporation of holistic assessment to accelerate the elicitation process in FITradeoff was recently proposed by de Almeida et al. [19]. In this context, this section describes the characteristics of FITradeoff and its adaptation to the negotiation context.

Recently, Frej et al. [10] developed an approach for multi-issue bilateral negotiation support based on the FITradeoff for preferences elicitation. Considering an additive aggregation model to evaluate alternatives (packages), the basic assumption of the FITradeoff method is that the values of criteria (issues) scaling constants are not exactly known by negotiators, and that they can only provide partial information about preferences.

The preferences elicitation process with the FITradeoff method begins with each negotiator declaring the ranking of issues according to their preferences, in descending order of relative importance. Then, the process continues based on an iterative question-and-answer process. These questions are designed to get negotiators to reflect on the tradeoffs between different levels of issues [10, 18, 19, 22]. Questions comparing hypothetical consequences considering tradeoffs between adjacent issues are posed to the negotiator [10]. Such comparisons are made interactively, for each pair of adjacent issue, following a heuristic originally proposed by de Almeida et al. [18]. In this context, it is important to highlight that in the FITradeoff model for negotiation support, the negotiator is not asked to directly evaluate the packages, but simply choose which one is best for him, according to his preferences.

The information obtained from the comparison of consequences are converted into inequalities, that act as constraints for Linear Programming models that run, at each interaction, trying to find dominance relations between candidate packages within the current weight space defined by the negotiator according to the information provided during the preference elicitation process [10]. A partial – or complete – ranking of such packages is built based on this dominance relations [10, 22]. In this ranking, negotiation packages are separated into different ranking positions according to the number of alternatives by which an alternative is dominated. The ranking may be partial due to the incomparability relationships that may arise, which is justified by the level of information provided by the negotiator. Due to the flexibility of the process, the user can, at any time, provide more preference information to obtain an increasingly refined ranking, as well as, at any time, decide to interrupt the preference elicitation process based only on the partial results identified.

2.2 Negotiation Protocol

The FITradeoff-based negotiation protocol [10] follows a sequential structure of steps, including the pre-negotiation, negotiation, and post-settlement phases. Initially, the input data of the negotiation process consists of the performance matrix of the negotiation problem considering all possible packages (obtained by combining the options of the issues, considering these as discrete), and the direction of preferences of the negotiators, whether minimization or of maximization on the options of the issues.

In the pre-negotiation phase, each negotiator ranks the issues scaling constants based on their preferences, in descending order of relative importance. At this point, a partial ranking of candidate packages is generated for each party, according to the preference information provided. It is important to highlight that each negotiator has his/her own ranking, but no preference information about the counterpart's ranking is shared, during the whole process.

Based on the ranking of packages obtained at this moment, the negotiator can decide whether or not he/she wants to propose an offer, starting the negotiation phase. If he is not ready to do it – which can be justified by the fact that the top positions in the ranking of packages have not yet been defined –, the negotiator can provide more information on preferences, aiming to obtain an even more refined ranking. As discussed in the previous section, this preference information is given by answering preference questions between hypothetical packages imposed by the FITradeoff method. As more information is provided by the negotiator, more refined is the ranking of packages. The negotiator can continue to answer the questions proposed by the method until he feels comfortable proposing the first offer based on the current ranking of packages.

When conducting the negotiation, the negotiator can then consult the current ranking of packages as a resource to aid in the formulation of an offer, as well as in the analysis of an offer made by his/her counterpart. That is, based on their ranking of packages, the negotiator is able to visualize in which position an offer he wants to make is found, as well as an offer proposed by his counterpart. Therefore, if the negotiator has not completed the entire preference elicitation process, he has the flexibility of being able to carry out, in parallel, the elicitation of preferences and the exchange of offers/packages until an agreement is identified. In this way, it is not necessary to complete the entire preference elicitation process only in the pre-negotiation phase.

The model proposed by Frej et al. (2021) works based on a dynamic set (or subset of remaining packages) of packages, in the sense that it starts with the entire set of possible packets and, during the negotiation, for some reason, they are eliminated from the negotiation. More specifically, according to Frej et al. (2021) a given package will be eliminated from the negotiation process if one of the following conditions is true. The first condition: A package P is proposed by one of the parties, but is rejected by the counterpart. The second condition: a negotiator proposes an offer; this offer is declined by the counterpart; and package P is dominated by this offer in the counterpart ranking. In other words, for each offer package made by a negotiator and rejected by the counterpart, this offer package, as well as the packages that are dominated by it in the ranking of counterpart, are eliminated from the negotiation, as it implies that these packages are less preferable for the negotiator who received the offer as they are at lower levels than the offer package in their ranking. Thus, at each round of negotiation, the packages that are no longer preferable to the parties are then discarded from the process. This mechanism is useful to ensure convergence towards an agreement that is reasonable for both parties, so that by eliminating unwanted packages, negotiators can reach an agreement more quickly [10].

Finally, once a package/agreement P* has been chosen by the negotiators, the last phase of the negotiation process begins, the post-settlement phase. At this phase, Pareto optimization is performed in order to verify whether the package chosen by the parties was in fact a solution optimal Pareto. More precisely, a joint improvement analysis is considered, which consists of verifying if there is a package, from the initial set of packages, better than P* for both parties. According to Frej et al. [10], this is done by calculating the ranking of all packages for both negotiators, according to the preference information they provided throughout the process. Therefore, if there is a package that is ranked better than P*, in both rankings of negotiators, then P* is not Pareto optimal, and so this other solution identified – which is better than P* for both negotiators – is presented to them, leaving it up to them to decide whether or not to accept the recommendation. Otherwise – i.e., if there is no package that ranks better than P* in both rankings – then P* is a Pareto optimal solution to the negotiation problem and the process is finalized.

3 Simulation Experiment

3.1 Description of the Simulation Experiment

In order to better analyze the behavior and particular aspects of the FITradeoff-based negotiation protocol, a simulation experiment was organized involving bilateral negotiations, which were conducted in FITradeoff electronic negotiation system. The purpose of such experiments was to investigating the effects of the elicitation process with the use of partial information, flexible and interactive for decision-making in negotiation problems with multiple issues.

For the development of the simulation experiment, a sample of 22 postgraduate students in production engineering from a university in Brazil was selected to act as negotiators. The students were then divided into pairs, to participate in a negotiation problem anonymously and randomly, totaling 11 pairs and, respectively, 11 different negotiation simulations. Thus, negotiators using the electronic platform FITradeoff for Negotiation, can test their knowledge in relation to negotiation, as well as the usability of the platform. In this simulation experiment, for all pairs, the same negotiation problem was used, which is one of the classic problems used in studies with Inspire [12], that concerns the signing of a contract between two agents, the which will be presented in Sect. 3.2.

Furthermore, detailed private information was provided for each participant that specifies the directions of preferences in relation to the negotiation issues, more precisely, whether the objective for each issue is to maximize or minimize and the preferential rank of the scaling constants.

3.2 Negotiation Problem

The problem used in this simulation experiment is an Inspire standard commercial negotiation problem [12] which is related to the signing of a contract at WorldMusic between two agents representing: an entertainment company, represented by Mosico and a musician represented by Fado, which was also used in [8, 10]. The purpose of the negotiation is to establish an agreement between the negotiating parties so that the agreement reached is satisfactory for both the entertainment company and the musician. This negotiation involves four issues, each with a predefined set of options:

- Number of promotional concerts (5; 6; 7; 8);
- Number of new songs (11; 12; 13; 14; 15);
- Royalties for CDs (1.5%; 2.0%; 2.5%; 3.0%);
- Contract signing bonus, in monetary values expressed in '000s (125; 150; 200).

Table 1 presents the initial preference information considered for Mosico and Fado, which is the order that is used by each negotiator in Step 1 of the FITradeoff method and the preferences directions.

Table 1. Initial preference information for Mosico and Fado.

Issue	Order (Mosico)	Preference (Mosico)	Order (Fado)	Preference (Fado)
Concerts	1	Max	1	Min
Songs	2	Max	2	Max
Royaties	3	Min	4	Max
Contract signing bonus	4	Min	3	Max

It is important to point out that, in order to simplify this problem, in the simulation experiment, it was admitted that the preferences of both negotiators are monotonic.

In the protocol for eliciting preferences in negotiation with multiple issues based on the FITradeoff method proposed by Frej et al. [10], some assumptions are considered, among them is the assumption that the negotiators' preferences for the issues' options are strictly monotonic (either strictly increasing or strictly decreasing). This assumption comes out from the original FITradeoff method itself [18, 22], which assumes monotonicity for value functions. In situations where a certain criterion (issue) has indeed a non-monotonic behavior, there are alternative ways for dealing with this, as argued by Belton & Stewart [26]; for instance, it can be recommended to revise the set of criteria (issues) and possibly split the non-monotonic criterion into two criteria: a strict increasing monotonic criterion and a strict decreasing monotonic criterion [26].

Moreover, it should be pointed out that, to build the intracriteria value function, the value of an option on a given issue is normalized – scale transformation – on a linear scale from 0 to 1, where 0 is the value of the worst option of an issue according to the preferences of the negotiator, and 1 is the value of the best option of an issue according to the preferences of the negotiator. Intermediate options, on the other hand, receive a value between 0 and 1, linearly. The linear function was considered for simplification purposes; however, the FITradeoff method also allows non-linear value functions to be considered in intracriterion evaluation. For more details on other ways to obtain such intracriteria value functions, the work by Belton and Stewart [26] presents a better discussion on this topic.

From the set of options for each question, the possible combinations of negotiation packages result in a set of 240 packages, which is the initial set of negotiation alternatives to start the interactive process of exchanging offers between the negotiating parties.

In the simulations, each pair of participants behaved like Mosico and Fado, so that the information about the order of the issues for Step 1 of the FITradeoff method and the directions of preferences followed according to the model of this problem, being sent to the participants in advance, individually and anonymously. Thus, 11 students worked with the Mosico profile and 11 students worked with the Fado profile. This means that each participant received their preference information privately according to their profile and, consequently, did not have knowledge about their counterparts' information.

In the following section, the results obtained from the experiment with negotiation simulations are presented, elucidating the main characteristics of all the analyzes performed.

4   Results

Based on the simulation experiment of the contract signing problem between Mosico and Fado, initially, some variables were defined to carry out the analyzes and, consequently, a better understanding of the negotiators' behavior using the preference elicitation protocol. Table 2 presents the variables collected and used in the analyzes presented below.

Table 2. Variables analyzed.

Variable	Description
NON	Number of offers made in the negotiation
NQN1	Number of questions answered by negotiator 1
NQN2	Number of questions answered by negotiator 2
NEPo	Number of eliminated packages by the offer
nEP (%)	% Eliminated of packages
NRP	Number of remaining packages
nRP (%)	% Number of remaining packages
NBPOP	Number of better packages after Pareto optimization

From the defined variables, a summary table of the results of the negotiation simulations between the participants was constructed, aiming to present a summary of the behavior of each pair of negotiators until an agreement was identified. Table 3 elucidates the summary of each negotiation, in which each line refers to a pair of participants simulating the problem of Mosico and Fado.

Table 3. Synthesis of simulations using the negotiation protocol with FITradeoff.

Variable	NON	NQN1	NQN2	NEPo	nEP (%)	NRP	nRP (%)	NBPOP
Pair 1	4	9	12	49	20%	191	80%	0
Pair 2	4	16	14	215	90%	25	10%	0
Pair 3	11	20	20	238	99%	2	1%	0
Pair 4	13	7	17	237	99%	3	1%	0
Pair 5	2	13	12	26	11%	214	89%	28
Pair 6	5	10	7	117	49%	123	51%	26
Pair 7	12	17	11	236	98%	4	2%	0
Pair 8	9	19	16	234	98%	6	2%	0
Pair 9	6	16	8	115	48%	125	52%	2
Pair 10	5	12	17	219	91%	21	9%	10
Pair 11	9	20	4	229	95%	11	5%	1

Based on the identified results shown in Table 3, a preliminary descriptive statistical analysis was performed, in order to withdraw broader conclusions about the 11 negotiation simulations. Table 4 presents the results in relation to the mean, median, maximum and minimum values and the amplitude for each analyzed variable in the sample. It is important to note that for the variables NON, NQN1, NQN2, NEPo, NRP and NBPOP the corresponding value for the mean was approximated to the largest integer, since these variables consist of integer values. It is worth mentioning that NRP + NEPo = 240 and nEP (%) + nRP (%) = 100 (%).

Table 4. Descriptive statistical analysis.

Variable	NON	NQN1	NQN2	NEPo	nEP (%)	NRP	nRP (%)	NBPOP
Mean	8	15	13	174	73%	66	27%	7
Median	6	16	12	219	91%	21	9%	0
Minimum	2	7	4	26	11%	2	1%	0
Maximum	13	20	20	238	99%	214	89%	28
Range	11	13	16	212	88%	212	88%	28

By analyzing Table 4, it is possible to perceive that the participants in the simulations exchanged between two and thirteen offers, and that on average, the pairs exchanged only 8 offers, which means that each participant made around 4 offers until an agreement was reached.

In relation to the number of questions answered in the elicitation process, it can be noted that the 22 participants answered between 4 and 20 questions proposed by the method, and that 6 negotiators preferred not to finish the elicitation process: three negotiators with the Mosico profile of the pairs 2, 4 and 7, and three negotiators with the Fado profile of the pairs 3, 5 and 7. That is, approximately 28% of the sample of participants chose not to answer all the questions proposed by the FITradeoff method, supporting only on partial results to aid in the process of exchanging offers. It was also noticed that the negotiator with the profile of Mosico answered more questions in the elicitation of preferences with FITradeoff than the negotiator with the profile of Fado, returning an average of 15 and 13 responses, respectively. This analysis can also be observed in Table 3, where in 7 cases out of 11, the negotiator with the Mosico profile answered more questions than the negotiator with the Fado profile, which only in 3 cases answered more questions.

It is also observed, in Table 4, that the minimum and maximum number of packages eliminated in the simulations consisted of 26 to 238, respectively, as well as the minimum and maximum number of remaining packages was between 2 and 214. For more, we have that, on average, 174 packages were eliminated from the negotiation until an agreement was found, which corresponds to approximately 73% of the total set of packages initially available for negotiation. In relation to the variable of remaining packages, it is noticed that, on average, there were 66 remaining packages, which leads to approximately 27% of the initial set of packages in the negotiation.

Now, according to Table 3, we have that four pairs managed to converge to a agreement with less than 50% of the packages eliminated or even with more than 60% of the remaining packages that could still be offered. Unlike these four pairs that had few packages eliminated during the negotiation to reach an agreement, the other seven pairs had more than 90% of the packages eliminated until an agreement was found.

With regards to the efficiency analysis of the agreement package, it was found that after calculating the Pareto optimization, on average, each pair of negotiators had at least 7 packages better than the one chosen by the participants. Furthermore, it was found that in 6 of the 11 pairs, approximately 55% of the simulated negotiations, the agreement package achieved by the parties was an efficient package; i.e., this means that, according to the negotiators' preferences, there was no package that simultaneously improve the result for both negotiators. Thus, for these 6 pairs, the identification package for each one was efficient, considering the preferences of both parties. However, for the other 5 pairs, or 45% of the simulation experiment negotiations, there was at least one package better than the package agreed upon after the efficiency analysis. In other words, in these five cases, there is one or more packages that return a better result of the negotiation simultaneously to the parties than the chosen agreement, being efficient recommendations and taking into account the elicited preferences.

5 Discussion and Conclusions

The main objective of this paper was to propose an analysis of the preference elicitation protocol using partial information in negotiation with multiple issues in order to assess the behavior of negotiators in their use, through a negotiation simulation experiment on an electronic negotiation system. The main characteristic identified of incorporating the FITradeoff compensatory method to elicit preferences in negotiation processes is the flexibility it provides to negotiators in the negotiation phases, this denotes a significant advantage when compared to other multi-criteria methods. More precisely, a preference elicitation procedure that is not restricted only to the pre-negotiation phase, but also to the negotiation conduction phase, allowing negotiators to elicit preferences and, in parallel, make offers based on their partial results for your counterpart.

In approach the FITradeoff for negotiation, negotiators do not have a single utility/multi-attribute value function to evaluate alternatives/packages, because exact weight values are not available. Thus, during the negotiation process, negotiators are no longer able to visualize a score for the offer and counteroffer, but based on the ranking of packages obtained by the mathematical algorithm on the FITradeoff method for the ranking problematic [22] (from the current stage of information provided by the negotiator), negotiators can make the decision on the building of their offers, as well as decide on accepting an offer from the counterpart. Thus, it is no longer necessary to present scores for the issues and the options of the issues in an unstructured way, where negotiators are required to provide complete information [1, 7, 12], assigning arbitrary scores to issues such as degree of importance, without them evaluating the ranges of issues options to formulate a utility function, as well as they don't present flexibility characteristics for negotiators.

From the results of the experiment with the negotiation simulations, the characteristic of flexibility to elicit preferences and, in parallel, make offers based on a partial ranking, as well as some negotiators do not need to complete the entire preference elicitation process until an agreement was identified were observed in the analyses.

During the negotiation simulations, for each rejected offer, the individual rankings of the negotiators were refined according to the elicitation protocol's packages elimination procedure [10], updating the candidate packages set. Thus, it was noticed that the elimination procedure of the dominated packages led to a convergence to the agreement more quickly in most simulations. This was confirmed based on the results presented in the previous section, in which negotiators 1 and 2 only needed to answer around 13 and 15 questions in the elicitation, respectively, and with only 8 offers and 174 packages eliminated, both on average, the pairs reached a mutual agreement that favored their preferences.

As seen in Sect. 4, in 5 of the 11 simulations, 6 participants did not need to complete the preference elicitation process and with only the partial results of the rankings was enough to support them in choosing the packages to be offered, and they certainly converged to an agreement in less time. It is important to highlight that, among these 6 participants, 3 had the Mosico profile and the other 3 had the Fado profile, distributed in 5 simulations, more precisely between the pairs 2, 3, 4, 5 and 7, and only the package chosen by pair 5 was not a Pareto optimal solution, as seen in Table 3 (NBPOP is different from zero for pair 5). However, for the others, the package chosen by them was Pareto optimal. This elucidates the potential of the elicitation protocol to structure and support negotiators in negotiations with multiple issues for a convergence to an effective agreement, where negotiators do not need to answer all the questions requested by the method until a complete preorder. Based on this finding, negotiators can visualize the partial results and, based on it, if they feel comfortable, make an offer or decide to accept a counteroffer from their counterpart even before completing the elicitation. These partial results refer to the visualization of the ranking of packages (partial or complete), obtained by the preference elicitation process, which can be a very useful tool to compare packages and can help negotiators in the decision-making process.

Still analyzing the results of the experiment, it was noticed that in 7 of the 11 simulations (pairs 2, 3, 4, 7, 8, 10, 11), more than 90% of the packages were eliminated from the initial set of the negotiation until they arrived to an agreement, and in the remaining 4 simulations, less than 50% of the packages were eliminated. It can be noted that for the subset of 14 participants, the simulation results show that these negotiations were more competitive, as the pairs eliminated more than 90% of the initial package set. More specifically, pairs 3, 4 and 7, where participants practically eliminated all packages from the negotiation until an agreement was found, leaving only 2, 3 and 4 packages remaining, respectively. This result raises hypotheses about the participant's profile in relation to the negotiation context, where possible correlations may exist, and which need to be better analyzed by statistical hypothesis tests in future studies regarding the type of negotiator's profile with the variables: number of responses in the elicitation process and number of offers made and number of packages eliminated. Furthermore, behavioral studies can also draw conclusions about these hypotheses listed here.

As could be observed, the vast majority of pairs that evolved more in the negotiation process, managed to have a large number of packages eliminated throughout the process, which shows a good performance of the proposed protocol in relation to the speed of convergence to an agreement package. In other words, as more offers were made, more packages not preferred by the parties were eliminated from the negotiation. Thus, the quick convergence of the method to an agreement package that satisfies both parties is evidenced, especially when the problem involves a very large number of issues and options to be dealt. Therefore, with only partial information about preferences, it was possible to find dominance relationships between packages and, consequently, help negotiators to identify an agreement. Thus, participants only need to provide the amount of information they want through strict preference statements, also allowing them to terminate the elicitation process at any time. Nevertheless, it would be an interesting topic for future studies to compare different approaches for negotiation support with regards to the effort made by negotiators for eliciting preferences.

Furthermore, as seen in Table 3, it was observed that after the Pareto optimality calculation, the package chosen by pairs 5, 6, 9, 10 and 11 was not a Pareto optimal solution. In other words, in 45% of the negotiations in the experiment, there was at least one package that dominates, simultaneously, in both rankings of the negotiators, the package agreed by each pair participating in the simulated negotiation problem. This fact can be justified, mainly, by the ranking of packages obtained by each negotiator based on their preferences, more precisely, in terms of preference information provided in the elicitation process, since the Pareto optimality analysis is carried out from negotiator rankings. In addition, it is also hypothesized that the negotiators, perhaps, adopted a more cooperative behavior in conducting the negotiation, especially in the build of offers, which may have led these pairs to an agreement that would not be a Pareto optimal solution. Although pairs 5, 6, 9, 10 and 11 did not identify an efficient agreement, the analysis of post-settlement of protocol proposed by Frej et al. [10] allowed the identification of one or more packages that simultaneously improve the outcome of the negotiation for the negotiators, elucidating how this new approach to the treatment of negotiations can favor the interested parties to efficient agreements.

Through this negotiation simulation experiment, it could be observed that the FITradeoff-based negotiation protocol enables negotiators to reach an agreement on an efficient package, without having to complete the entire standard elicitation procedure, only providing partial preference information. As a limitation of this work, the sample size is indicated, which made it impossible to carry out other more complex statistical analyses on the performance of the protocol using partial information on the preferences of the negotiators. For future studies, it would be interesting to increase the sample size in order to have more robust results, as well as to investigate how negotiators profiles can influence the course of the offer exchanges within such protocol.

Acknowledgments. The authors are grateful for the support received and the funding provided by the Pernambuco Science and Technology Support Foundation (FACEPE); Brazilian National Council of Technological and Scientific Development (CNPq) and Coordination for the Improvements of Higher Education Personnel (CAPES). This work was supported by the FACEPE [grant numbers 1244-3.08/19].

References

1. Wachowicz, T., Roszkowska, E.: Can holistic declaration of preferences improve a negotiation offer scoring system? Eur. J. Oper. Res. (2021). https://doi.org/10.1016/j.ejor.2021.10.008
2. Wachowicz, T.: Decision support in software supported negotiations. J. Bus. Econ. **11**(4), 576–597 (2010)
3. Raiffa, H., Richardson, J., Metcalfe, D.: Negotiation Analysis: The Science and Art of Collaborative Decision Making. Harvard University Press, Cambridge (2002)

4. Mustajoki, J., Hamalainen, R.P.: Web-HIPRE: global decision support by value tree and AHP analysis. INFOR J. **38**(4), 208–220 (2000)
5. Wachowicz, T.: Negotiation template evaluation with calibrated ELECTRE-TRI method. In: de Vreede, G.J. (eds.) Group Decision and Negotiations 2010b, The Center for Collaboration Science, University of Nebraska at Omaha, pp. 232–238 (2010)
6. Górecka, D., Roszkowska, E., Wachowicz, T.: The MARS approach in the verbal and holistic evaluation of the negotiation template. Group Decis. Negot. **25**(6), 1097–1136 (2016)
7. Wachowicz, T., Roszkowska, E.: Holistic preferences and prenegotiation preparation. In: Kilgour, D.M., Eden, C. (eds.) Handbook of Group Decision and Negotiation, pp. 255–289. Springer, Cham (2021). https://doi.org/10.1007/978-3-030-49629-6_64
8. Roszkowska, E., Wachowicz, T., Kersten, G.: Can the holistic preference elicitation be used to determine an accurate negotiation offer scoring system? A comparison of direct rating and UTASTAR techniques. In: Schoop, M., Kilgour, D.M. (eds.) GDN 2017. LNBIP, vol. 293, pp. 202–214. Springer, Cham (2017). https://doi.org/10.1007/978-3-319-63546-0_15
9. Roszkowska, E., Wachowicz, T.: Inaccuracy in defining preferences by the electronic negotiation system users. In: Kamiński, B., Kersten, G.E., Szapiro, T. (eds.) GDN 2015. LNBIP, vol. 218, pp. 131–143. Springer, Cham (2015)
10. Frej, E.A., Morais, D.C., de Almeida, A.T.: Negotiation support through interactive dominance relationship specifcation. Group Decis. Negot. (2021). https://doi.org/10.1007/s10726-021-09761-y
11. Kersten, G.E., Lai, H.: Negotiation support and E-negotiation systems: an overview. Group Decis. Negot. **16**(6), 553–586 (2007)
12. Kersten, G.E., Noronha, S.J.: WWW-based negotiation support: design, implementation, and use. Decis. Support Syst. **25**(2), 135–154 (1999)
13. Schoop, M., Jertila, A., List, T.: Negoisst: a negotiation support system for electronic business-to-business negotiations in e-commerce. Data Knowl. Eng. **47**(1), 371–401 (2003)
14. Schoop, M.: Negoisst: complex digital negotiation support. In: Kilgour, D.M., Eden, C. (eds.) Handbook of Group Decision and Negotiation, 2nd edn., pp. 1149–1168. Springer, Cham (2021)
15. Wachowicz, T.: NegoCalc: spreadsheet based negotiation support tool with even-swap analysis. In: Climaco, J., Kersten, G.E., Costa, J.P. (eds.) Group Decision and Negotiation 2008: Proceedings—Full Papers, pp 323–329. INESC, Coimbra (2008)
16. Keeney, R.L., Raiffa, H.: Decision Analysis with Multiple Conflicting Objectives. Wiley, New York (1976)
17. Salo, A., Hämäläinen, R.P.: Preference assessment by imprecise ratio statements. Oper. Res. **40**(6), 1053–1061 (1992)
18. de Almeida, A.T., Almeida, J.A., Costa, A.P.C.S., Almeida-Filho, A.T.: A new method for elicitation of criteria weights in additive models: flexible and interactive tradeoff. Eur. J. Oper. Res. **250**(1), 17–191 (2016)
19. de Almeida, A.T., Frej, E.A., Roselli, L.R.P.: Combining holistic and decomposition paradigms in preference modeling with the flexibility of FITradeoff. CEJOR **29**(1), 7–47 (2021). https://doi.org/10.1007/s10100-020-00728-z
20. Sarabando, P., Dias, L.C., Vetschera, R.: Mediation with incomplete information: approaches to suggest potential agreements. Group Decis. Negot. **22**(3), 561–597 (2013)
21. Weber, M., Borcherding, K.: Behavioral influences on weight judgments in multiattribute decision-making. Eur. J. Oper. Res. **67**(1), 1–12 (1993)
22. Frej, E.A., de Almeida, A.T., Costa, A.P.C.S.: Using data visualization for ranking alternatives with partial information and interactive tradeoff elicitation. Oper. Res. Int. J. **19**(4), 909–931 (2019). https://doi.org/10.1007/s12351-018-00444-2
23. Kang, T.H.A., Frej, E.A., de Almeida, A.T.: Flexible and interactive tradeoff elicitation for multicriteria sorting problems. Asia Pac. J. Oper. Res. **37**(5), 2050020 (2020)

24. Frej, E.A., Ekel, P., de Almeida, A.T.: A benefit-to-cost ratio based approach for portfolio selection under multiple criteria with incomplete preference information. Inform. Scien. **545**(4), 487–498 (2021)
25. Borcherding, K., Eppel, T., Von Winterfeldt, D.: Comparison of weighting judgments in multiattribute utility measurement. Manage. Sci. **37**(12), 1603–1619 (1991)
26. Belton, V., Stewart, T.: Multiple Criteria Decision Analysis: An Integrated Approach. Springer, New York (2002). https://doi.org/10.1007/978-1-4615-1495-4

Multi-agent Task Allocation Under Unrestricted Environments

Takahiro Suzuki[(✉)] [iD] and Masahide Horita[iD]

Department of Civil Engineering, The University of Tokyo,
7-3-1, Hongo, Bunkyo-ku, Tokyo, Japan
{suzuki-tkenmgt,horita}@g.ecc.u-tokyo.ac.jp

Abstract. Suppose that a construction manager is assigning agents (construction robots) to the set of tasks M. Each task has a weak/linear preference over the coalitions of robots. However, the manager only knows the preferences of $N \subseteq M$; perhaps because estimating the preferences of $M \setminus N$ takes an unreasonable amount of time/cost. The present paper explores whether the manager can find a Pareto optimal (PO) allocation of the robots for the entire M. Two approaches are axiomatically studied. One approach is to find an entire allocation that is PO under any realization of the preferences of $M \setminus N$. The other is to first allocate within the tasks in N and then assign the remaining robots within $M \setminus N$ (after their preferences are obtained), so that the entire allocation is PO. The contribution of this paper is twofold. We first prove that the first (second) approach is possible if and only if there exists an allocation for N that is PO and non-idling (NI) (weakly non-idling [WNI]); where NI is an axiom demanding that no allocation weakly dominates the allocation with some agents unassigned. The second result is from an algorithmic perspective; we prove that serial dictatorship must find a PO and WNI allocation (if it exists) under a linear preference domain.

Keywords: Pareto optimality · Multi-robot task allocation · Serial dictatorship

1 Introduction

How can construction robots effectively cooperate when executing a construction project in a completely new environment such as the moon? Consider a moon project where several robots (i.e., construction machines such as backhoes and dump trucks) cooperate in order to excavate soil from hills, moving the soil to a disposal site. In such a case, what is the best allocation of robots on each hill? One of the major difficulties could be a lack of similar experiences in the past, which causes great uncertainty regarding the nature of the tasks. We call this an unrestricted environment (UE). Indeed, some hills may be too compacted to be excavated by some backhoes; meanwhile, at the other hills, the work yard might be highly confined. Such characteristics are often unknown until the project begins, to some extent.

The present paper explores how to assign multiple agents (specifically, construction robots) to different tasks in a UE. Let M be the set of all tasks (e.g., soil hills). Each

D. C. Morais and L. Fang (Eds.): GDN 2022, LNBIP 454, pp. 31–43, 2022.
https://doi.org/10.1007/978-3-031-07996-2_3

task has a weak/linear preference on the coalitions of robots ("coalition C is preferred to D in task i's preference" means that C is better than D in executing the task i). The construction manager must assign robots to each task,[1] knowing only the preferences of $N (\subseteq M)$. The preferences of tasks in $M \setminus N$ are assumed to be unknown (this is because of a lack of experience, the unreasonable cost/time that is required to determine the preferences, etc.) when the manager determines the assignments. The preferences might only be known later, perhaps because sufficient information for estimating their preferences would be achieved as progress is made.

Our central problem pertains to how the manager can reach a Pareto optimal (PO) allocation for the entire set of tasks M.[2] Ideally, it is preferable if one can initially find a PO allocation X for M. However, the existence of the unknown preferences often makes this attempt difficult; it is often the case that X is PO under some preference profiles of $M \setminus N$, but is not PO under others. Is it ever possible to find the PO allocation with the existence of unknown tasks? If there is a known task with a strictly monotonic preference (more robots must always fare well), then the answer is trivially yes—assigning all robots to such a task is surely PO. Such a monotonic preference looks natural when economic objects are assigned. However, it might not be the case when the robots are assigned to each task; this is because excessive input can cause robots to unnecessarily occupy the work yard, resulting in low productivity for the entire project.

The above observation indicates that some preference profiles of N are sufficient to reach a PO allocation for the entire M; however, this is not always the case (this fact will be detailed formally in both Example 1 and Example 2, which are found in subsequent sections). This paper is an axiomatic and algorithmic study of such sufficiency and proceeds as follows. Section 3 introduces the basic allocation model. Section 4 formulates the above observations and characterizes sufficiency by the existence of an allocation for N that is PO and non-idling (NI)—which is an axiom demanding that no allocation can weakly dominate with some robots unassigned (Proposition 2). Section 5 stands on the algorithmic part and argues how one can reach such PO allocation for all tasks M. Finally, Sect. 6 includes concluding remarks.

2 Related Literature and Our Contribution

Based on an approach using social choice theory, this paper conducts an axiomatic study of PO allocations with limited preferential information. In general, the allocation of indivisible objects is an interdisciplinary topic (cf., [1, 2] for a general survey) with a vast number of applications (the indivisible goods can be houses [3], agents to be assigned to group activities [4, 5], robots to be assigned to tasks [6], etc.) and research topics (fair allocations [7, 8], strategyproofness [9], computational complexity of allocation procedures, etc.). Among these, the following three topics are the closest to our models.

---

[1] As noted in this paragraph, the present paper mainly argues the context of assigning robots between tasks. Substituting tasks with people and robots with indivisible objects, one can translate our argument into the standard allocation models.

[2] Postponing the assignment for $M \setminus N$ (until their preferences become clear) is another possible approach (further discussion will be made after Definition 3). However, this might lead to the delay of the project due to the amount of uncertainties.

(1) *Many-objects-to-one-agent allocation*: Our model can be seen as an application of the so-called many-objects-to-one-agent allocation model (without monetary transfer) [9–11]. Extending the classic allocation model where each person can receive— at most—one object, Pápai [9] studies an allocation model where each agent has linear preferences over the power set of the set of objects; furthermore, each agent can be assigned to more than one object. Then, Pápai characterizes sequential dictatorship (serial dictatorship) as the unique allocation rule being strategyproof, nonbossy (totally nonbossy), and PO. Further generalizations, when agents have responsive (and separable) preferences, are discussed in other studies [10, 11]. In our terminology, the allocation rule is interpreted as a procedure that assigns a coalition of robots to each task. What distinguishes our model from most others is that we deal with uncertainty with respect to the preferences of some tasks (i.e., preferences of tasks are only partly known). Section 5 sheds some light on the property of serial dictatorship in such a situation. While the above-mentioned research indicates a certain connection between serial dictatorship and PO, the connection can disappear under certain conditions (e.g., when each agent is to be assigned a fixed number of objects [12]). As a similar result, Sect. 5 also argues that serial dictatorship can fail to yield a weakly non-idling (WNI) and PO allocation—even when such an allocation exists.

(2) *Matching with partially known preferences*: From an informational point of view, our model is an application of the two-sided stable matching model with partial information on people's preferences [13, 14]. Note that our model is one-sided in that only tasks are assumed to have preferences, whereas robots do not. When the two-sided matching market (e.g., firms and candidates, or hospitals and residents) becomes large, it often becomes impractical to expect that each player has a complete preference over all the others. In such a case, interviews between players often form a key role in eliciting the strict preferences. However, this itself can be both a time- and cost-consuming activity that should be avoided as much as possible. Focusing on such difficulty, [13] proposed a novel model for two-sided matchings; where players begin with partial preferences of the others and study how to minimize the number of interviews that are required to reach a stable matching. A growing number of studies further investigate such environments (e.g., unified model with preference elicitation [14], several uncertainty models [15], etc.). Our model applies this lack of preference information to a one-sided task allocation between multiple robots.

(3) *Multi-robot task allocation* (*MRTA*): This paper is connected with multi-robot task allocation (MRTA) in the field of robotics engineering. Using the taxonomy of the MRTA problem, which is proposed by [6], our model can be expressed as 1) single-task robots (ST) (i.e., each robot can execute, at most, one task); 2) multi-robot tasks (MR) (i.e., some tasks can require multiple robots); and 3) instantaneous assignment (IA). One major difference between our model and such ST-MR-IA problems in [6] is that our model stands on purely ordinal information (i.e., it does not use the cardinal utilities of each task). This enables us to naturally apply social choice theory into cases where cardinal performance is difficult to measure.

3 Basic Model: Allocation of Robots

Let K be a nonempty set of robots $(3 \leq |K| < +\infty)$. Let $\mathcal{K} = 2^K$ be the set of all coalitions. For a set N of tasks $(2 \leq |N| < +\infty)$, an N-allocation is a function $X : N \to \mathcal{K}$—such that for all $i, j \in N$,

$$X(i) \cap X(j) = \phi \text{ whenever } i \neq j. \tag{1}$$

From now on, for the ease of notation, $X(i)$ is denoted by $X_i{}^3$. This is interpreted as the set of robots that are assigned to task i. Equation (1) means that no robot is assigned to more than one task. Note that some robots can be unassigned (i.e., there may be some $r \in K$, such that $r \notin X_i$ for all $i \in N$). The set of all N-allocations is denoted by $\mathcal{X}_N$. Each task $i \in N$ has a *preference* $\succsim_i$ over $\mathcal{K}$, which is either a weak order or a linear order over $\mathcal{K}$. We often denote by $\succsim_A$ the preference profile of the set of tasks A. Task i's preference $\succsim_i$ represents which coalition is best suited to the task (i.e., for task $i \in N$ and coalitions $C, D \subseteq K$, $C \succsim_i D$, it means that the coalition C is at least as good as D in executing the task i). As usual, we denote by $\succ_t$ (resp. $\sim_t$) the asymmetric (resp. symmetric) part of $\succsim_t$. The set of all admissible preferences is denoted by $\mathcal{D}$, called the *domain* (*of preferences*). In this paper, we assume that the domain is either *general domain* $\mathcal{D} = \mathcal{R}(\mathcal{K})$ (which is made up of all the weak orders over $\mathcal{K}$) or *linear domain* $\mathcal{D} = \mathcal{L}(\mathcal{K})$ (which is made up of only the linear orders over $\mathcal{K}$). Subsequent arguments hold under either domain, unless specified. A list of preferences for the set N of tasks is called an *N-profile*, which designates an element of $\mathcal{D}$ to each task in N. The set of all N-profiles under domain $\mathcal{D}$ is denoted by $\mathcal{D}_N$. An N-allocation rule is a mapping $f : \mathcal{D}_N \to \mathcal{X}_N$ that maps each N-profile to an N-allocation. As was previously noted in Sect. 1, our model is very close to the standard many-objects-to-one-person model [9–11]. Indeed, when $\mathcal{D} = \mathcal{L}(\mathcal{K})$, our N-allocation rule is equivalent to the allocation rule in [9] (robots and tasks in our model correspond with indivisible goods and people/agents, respectively). Nevertheless, our model is distinguished from these in that we consider various sets of tasks; this is why we state an N-allocation rule rather than simply an allocation rule.

Let X and Y be N-allocations. We say that Y *weakly* (*Pareto*) *dominates* X with respect to $\succsim = (\succsim_i)_{i \in N} \in \mathcal{D}_N$ if $Y_i \succsim_i X_i$ for all $i \in N$. Also, Y (*Pareto*) *dominates* X with respect to $\succsim = (\succsim_i)_{i \in N} \in \mathcal{D}_N$ if Y weakly dominates X and $Y_j \succ_j X_j$ for some $j \in N$. The properties of allocations, as well as allocation rules, are defined in order.

Definition 1 (Properties of allocations)
An N-allocation $X \in \mathcal{X}_N$ is said to be the following:

– *PO* with respect to $\succsim \in \mathcal{D}_N$ if there is no other N-allocation $Y \in \mathcal{X}_N$ that dominates X.
– *Non-idling* (*NI*) with respect to $\succsim \in \mathcal{D}_N$ if for all $Y \in \mathcal{X}_N$,

$$\left(Y \text{ weakly dominates } X \text{ w.r.t. } \succsim \right) \Rightarrow \bigcup_{i \in N} Y_i = K.$$

---

[3] With a little abuse of notation, we sometimes regard an N-allocation X as an $|N|$-tuple of subsets of K—satisfying (1) rather than the function from N to $\mathcal{K}$ literally.

– *Weakly Non-idling (WNI)* with respect to $\succsim \in \mathcal{D}_N$ if for all $Y \in \mathcal{X}_N$,

$$(Y \text{ weakly dominates } X \text{ w.r.t. } \succsim) \Rightarrow \bigcup_{i \in N} Y_i \supseteq \bigcup_{i \in N} X_i.$$

PO is a standard axiom. An N-allocation X is PO if, and only if, there is no N-allocation that dominates X. To the best of authors' knowledge, the other two properties are new. An N-allocation X is called NI if X is weakly dominated by another N-allocation $Y \in \mathcal{X}_N$ only if such Y makes use of all the robots ($\bigcup_{i \in N} Y_i = K$). In other words, no allocation can weakly dominate X with some robots unassigned. Similarly, $X \in \mathcal{X}_N$ is called WNI if X is weakly dominated by another $Y \in \mathcal{X}_N$ only if such Y makes use of *more robots than* X ($\bigcup_{i \in N} Y_i \supseteq \bigcup_{i \in N} X_i$). NI implies WNI; however, the converse is generally not true. These axioms rule out the redundant use of robots; furthermore, they turn out to be key axioms for PO allocations (Proposition 2 and 3).

Definition 2 (Properties of allocation rules)
An N-allocation rule $f : \mathcal{D}_N \to \mathcal{X}_N$ is said to be the following:
– *PO* if, for any profile $\succsim \in \mathcal{D}_N, f(\succsim)$, is PO.
– *Non-idling (NI)* if for all profiles $\succsim \in \mathcal{D}_N$, if there exists at least one N-allocation that is NI under $\succsim$, then $f(\succsim)$ is one of such allocations.
– *Weak Non-idling (WNI)*, if for all profiles $\succsim \in \mathcal{D}_N$; if there exists at least one N-allocation that is WNI under $\succsim$, then $f(\succsim)$ is one of such allocations.

4 Allocation with Information at Hand

Situation Caused by the UE. Let $M = \{1, 2, \cdots, m\}$ be the set of all tasks and $\succsim = (\succsim_1, \succsim_2, \cdots, \succsim_m) \in \mathcal{D}_M$ be an M-profile. Assume that the construction manager knows only the preferences of the first n tasks, $N = \{1, 2, \cdots, n\}$ ($n \leq m$). Is it possible to find a PO M-allocation based only on the information at hand (i.e., $\succsim_1, \succsim_2, \cdots, \succsim_n$)?

Observation 1 (The case of $n = m$)
When $n = m$ (i.e., when $\succsim$ is completely known), the answer is always yes. The Pareto dominance relation is surely both a transitive and antisymmetric binary relation over $\mathcal{X}_M$. With $\mathcal{X}_M$ as a finite set, there must exist some maximal element—which is a PO M-allocation. Note that such PO M-allocations are, in general, not unique.

On the other hand, when $n < m$ (i.e., the preference of some tasks are unknown), the answer depends on what $\succsim_1, \succsim_2, \cdots, \succsim_n$ are like (see Example 1 and Example 2).

A preference $\succsim$ is called *monotonic (with respect to set inclusion)* if for all $C, D \in \mathcal{K}$, $(C \supseteq D \Rightarrow C \succsim D)$. A preference $\succsim$ is *strictly monotonic* if it is monotonic and for all $C, D \in \mathcal{K}, (C \supset D \Rightarrow C \succ D)$. Similarly, *anti-monotonic* and *strictly anti-monotonic* preferences are defined by reversing $\succsim$ and $\succ$—respectively (i.e., $\succsim$ is *anti-monotonic* if for all $C, D \in \mathcal{K}, [C \supseteq D \Rightarrow D \succsim C]$ and *strictly anti-monotonic* if it is anti-monotonic and for all $C, D \in \mathcal{K}, [C \supset D \Rightarrow D \succ C]$). In the study of preferences over sets, the monotonicity with respect to set inclusion is argued in [16, 17].

Example 1 (A task with strictly monotonic preferences)
Let $M = \{1, 2\} \supset \{1\} = N$ and $K = \{b_1, b_2, d_1, d_2\}$ (for $i = 1, 2, b_i$ is interpreted

as the ith backhoe and d_i the ith dump truck). Regard tasks 1 and 2 as two hills of soil from which the robots must cooperatively excavate soil and move it to a disposal site. Suppose that task 1's preference $\succsim_1$ is strictly monotonic (i.e., the addition of an extra robot must always fare well). In such a case, the M-allocation $(X_1, X_2) = (K, \phi)$ is PO under any $\succsim_2 \in \mathcal{R}(\mathcal{K})$.

Example 2 (A case where PO allocation cannot be determined)
With the same M, N, K, consider the following preferences: for all $C, D \in \mathcal{K}$, (i) if $b_2 \notin C$ and $b_2 \in D$, then $C \succ_1 D$, (ii) Suppose either $b_2 \in C \cap D$ or $b_2 \notin C \cup D$: if $\{b_1, d_1\}$ is in C—but not in D—then $C \succ_1 D$ and (iii) otherwise, $C \sim_1 D$. An interpretation of this is as follows. Only the first backhoe b_1 can excavate the soil from task 1 (perhaps because the first hill is too compacted to excavate for b_2), and only the first dump truck d_1 is compatible with b_1 (perhaps due in part to the size of the machines). Furthermore, the assignment of backhoe b_2 decreases the productivity. In this case, any M-allocation $X = (X_1, X_2)$ can fail to be PO under certain $\succsim_2$. First, assigning b_2 to task 1 is certainly inefficient (by ruling out b_2, the allocation improves in a Pareto sense). Therefore, assume $b_2 \notin X_1$. If $b_2 \in X_2$, X is certainly not PO when $\succsim_2 = \succsim_1$. Otherwise (i.e., when $b_2 \notin X_2$), X is not PO either, if $\succsim_2$ is strictly monotonic (in that case, the addition of b_2 to task 2 is Pareto-improving).

These two examples illustrate the difficulty of finding a PO allocation under the UE. When the preferences are only partly known (and especially when non-monotonic preferences matter), it is possible that for any M-allocation (global allocation) X, X is not PO under a certain realization of preferences of $M - N$. The possibility of such non-monotonic preferences is supposed to be typical of our construction contexts, where an excessive quantity of allocations to a task can ruin the efficiency—perhaps because they waste either the workspace or resources. Proposition 1 is the formal description of this observation.

Proposition 1
Let $N = \{1, 2, \cdots, n\} \subset M = \{1, 2, \cdots, m\}$ with $n < m$.
(i) There exists $\succsim_N \in \mathcal{D}_N$; such that for all $X \in \mathcal{X}_M$, there exists $\succsim_{M-N} \in \mathcal{D}_{M-N}$ such that X is not PO under $\succsim = (\succsim_N, \succsim_{M-N})$.
(ii) If there exists a task in N with a strictly monotonic preference, then there exists an M-allocation X such that for all $\succsim_{M-N} \in \mathcal{D}_{M-N}$, X is PO under $\succsim = (\succsim_N, \succsim_{M-N})$.

Proof of Proposition 1:
(i) Suppose that each task $i \in N$ ranks C_i as the best coalition (i.e., $C_i \succ_i D$ for all $D \neq C_i$); furthermore, suppose that $C_1, C_2, \cdots, C_n$ is mutually disjoint and yet collectively non-exhaustive (i.e., $i \neq j \Rightarrow C_i \cap C_j = \phi$, and $C_1 \cup C_2 \cup \cdots \cup C_n \neq K$). A typical example is when each $\succsim_1, \succsim_2, \cdots, \succsim_n$ is strictly anti-monotonic. Let $K' := K \backslash (C_1 \cup C_2 \cup \cdots \cup C_n)$. Fix any $X \in \mathcal{X}_M$. If $X_m \neq \phi$, X is dominated by $Y = (C_1, C_2, \cdots, C_n, \phi, \cdots, \phi)$ (assigning C_i to $i \in N$ and no robots to tasks in $M \backslash N$) when every $\succsim_{n+1}, \cdots, \succsim_m$ is strictly anti-monotonic. Similarly, if $X_m = \phi$, X is dominated by $Y' = (C_1, C_2, \cdots, C_n \phi, \cdots, \phi, K')$ (modifying Y by assigning

K' instead of ϕ to task m) when $\succsim_{n+1}, \cdots, \succsim_{m-1}$ is strictly anti-monotonic and $\succsim_m$ is strictly monotonic. In each case, X is not PO under a certain profile.

(ii) When $j \in N$ has a strictly monotonic preference, $X \in \mathcal{X}_M$ such that $X_j = K$ and $X_i = \phi$ ($i \neq j$) is surely PO. ∎

Proposition 1 motivates us to specify the class of N-profiles with which one can find a PO M-allocation. Let us say that an M-allocation $X \in \mathcal{X}_M$ is *absolutely Pareto optimal* (*APO*), with respect to $\succsim_N \in \mathcal{D}_N$ if for all $\succsim_{M-N}$, X is PO under $\succsim = (\succsim_N, \succsim_{M-N})$. In addition, we say that an N-allocation $X \in \mathcal{X}_N$ is *weak-absolutely Pareto optimal* (*WAPO*), with respect to $\succsim_N$ (and M) if for all $\succsim_{M-N}$ there exists $X_{M-N} \in \mathcal{X}_{M-N}$ such that $(X, X_{M-N}) \in \mathcal{X}_M$ and it is PO under $(\succsim_N, \succsim_{M-N})$.

Definition 3 (Sufficient)
Let $N \subseteq M$. A N-profile $\succsim_N \in \mathcal{D}_N$ is said to be the following:
(1) *sufficient* if there exists an M-allocation $X \in \mathcal{X}_M$ that is APO, with respect to $\succsim_N$.
(2) *weakly sufficient* if there exists an N-allocation $X \in \mathcal{X}_N$ that is WAPO, with respect to $\succsim_N$ and M.

Sufficiency and weak sufficiency describe two different ways for reaching PO M-allocations in a UE. If $\succsim_N$ is sufficient, one can immediately find a PO M-allocation X without knowing the actual preferences of unknown tasks $M - N$. If $\succsim_N$ is weakly sufficient, on the other hand, then the manager can first assign the robots to well-known tasks N (i.e., to determine $X_1, X_2, \cdots, X_n$) and then assign the remaining robots to the unknown tasks $M - N$, so that the entire allocation becomes PO (i.e., to determine the other $X_{n+1}, \cdots, X_m$ so that $(X_1, X_2, \cdots, X_m)$ is PO) once the manager gets to know the preferences $M - N$. If $\succsim_N$ is not even weakly sufficient, then this two-step algorithm can potentially fail to work. It might be necessary to re-arrange $X_1, X_2, \cdots, X_n$ into $X_1', X_2', \cdots, X_n'$ in order to obtain a PO M-allocation. In other words, weak sufficiency considers how one can avoid re-arrangement in the two-step assignment (i.e., first, allocate the robots within N; then, allocate the remaining robots within $M - N$). Recall the earthmoving tasks seen in Example 1 and Example 2. Once a backhoe is assigned to a hill and it begins to excavate the soil, it would require extra time and funds to interrupt its current task and assign/move it to a new hill. Weak sufficiency demands that such an interruption is unnecessary to obtain an M-allocation.

A little reflection will tell us that sufficiency implies weak sufficiency, but not vice versa. The profile in Example 2 is not sufficient (essentially, this fact is already argued in the example itself), but it is weakly sufficient. To verify the latter, let $X_1 = \{b_1, d_1\}$; furthermore, let X_2 be the best element among $E := \{C \subseteq K \mid C \cap \{b_1, d_1\} = \phi\}$, with respect to $\succsim_2$ (with $\phi \in E$, such a X_2 must exist). Then, (X_1, X_2) is PO.

Proposition 2: (Weak) Sufficiency $\Leftrightarrow$ ∃ allocation being PO and (Weak) NI
Let $N = \{1, 2, \cdots, n\} \subset M = \{1, 2, \cdots, m\}$ with $n < m$.
(1) An N-profile $\succsim_N \in \mathcal{D}_N$ is sufficient if, and only if, there exists an N-allocation X that is PO and NI—with respect to $\succsim_N$.
(2) An N-profile $\succsim_N \in \mathcal{D}_N$ is weakly sufficient if, and only if, there exists an N-allocation X that is PO and WNI—with respect to $\succsim_N$.

Proposition 3: (W)APO ⇔ PO and (W)NI

Let $N = \{1, 2, \cdots, n\} \subset M = \{1, 2, \cdots, m\}$ with $n < m$.

(i) $\overline{X} = (X_1, X_2, \cdots, X_m) \in \mathcal{X}_M$ is APO at $\succsim_N \in \mathcal{D}_N$ if, and only if, $X = (X_1, X_2, \cdots, X_n) \in \mathcal{X}_N$ is PO and NI at $\succsim_N$ and $X_{n+1} = \cdots = X_m = \phi$.

(ii) $X = (X_1, X_2, \cdots, X_n) \in \mathcal{X}_N$ is WAPO with respect to $\succsim_N \in \mathcal{D}_N$ and M if, and only if, X is PO and WNI at $\succsim_N$.

Proposition 2 is easily obtained from Proposition 3. Thus, we prove only the latter.

Proof of Proposition 3:

[If part of (i)] Suppose $X = (X_1, \cdots, X_n) \in \mathcal{X}_N$ is PO and NI at $\succsim_N \in \mathcal{D}_N$. We prove that $\overline{X} := (X_1, \cdots, X_n, X_{n+1}, \cdots X_m) \in \mathcal{X}_M$, where $X_j = \phi$ for all $j \in M \backslash N$, is PO at any M-profile of the form $(\succsim_N, \cdot)$. Suppose not. Then, there exists $\succsim_{M-N}$ and $\overline{Y} = (Y_1, \cdots, Y_m) \in \mathcal{X}_M$ that dominates $\overline{X}$ at $(\succsim_N, \succsim_{M-N})$. This implies that $Y := (Y_1, \cdots, Y_n) \in \mathcal{X}_N$ weakly dominates X at $\succsim_N$. NI demands that $Y_1 + \cdots + Y_n = K$. As a result, $Y_j = \phi$ for all $j \in M \backslash N$. Now, we can conclude that Y must dominate X (otherwise, $\overline{Y}$ cannot dominate $\overline{X}$). This contradicts the assumption that X is PO.

[Only if part of (i)]. Suppose that $\overline{X} = (X_1, X_2, \cdots, X_m) \in \mathcal{X}_M$ is APO at $\succsim_N \in \mathcal{D}_N$. Let $X := (X_1, \cdots, X_n) \in \mathcal{X}_N$ (i.e., the first n elements of $\overline{X}$). We prove that (a) $X_j = \phi$ for all $j \in M \backslash N$, (b) X is PO, and (c) X is NI.

Proof of (a): Suppose, to the contrary, that $X_j \neq \phi$ for some $j \in M \backslash N$. Let $\succsim_{M-N} \in \mathcal{D}_{M-N}$ be strictly anti-monotonic. Then, $\overline{X}$ must be dominated by $\overline{Y} = (X_1, \cdots, X_n, \phi, \cdots, \phi)$ (i.e., for tasks in N, $\overline{Y}$ assigns the same as $\overline{X}$; for tasks in $M - N$, $\overline{Y}$ assigns nothing) at $(\succsim_N, \succsim_{M-N})$. This contradicts the assumption that $\overline{X}$ is APO.

Proof of (b): Suppose, to the contrary, that X is not PO. Then, there exists $Y = (Y_1, \cdots, Y_n) \in \mathcal{X}_N$ that dominates X. By this, as well as (a), it follows that $\overline{Y} = (Y_1, \cdots, Y_n, \phi, \cdots, \phi)$ (i.e., for tasks in N, $\overline{Y}$ assigns the same as Y; for tasks in $M - N$, $\overline{Y}$ assigns nothing) dominates $\overline{X}$. This contradicts the assumption that $\overline{X}$ is APO.

Proof of (c): Suppose that $Y = (Y_1, \cdots, Y_n) \in \mathcal{X}_N$ weakly dominates X. By (b), it must be that $Y_i \sim_i X_i$ for all $i \in N$. Contrarily, suppose that $Y_0 := Y_1 + Y_2 + \cdots + Y_n \neq K$. Then, $\overline{Y} = (Y_1, \cdots, Y_n, K - Y_0, \phi, \cdots, \phi) \in \mathcal{X}_M$ (assigning $K - Y_0$ to task $n + 1$ and ϕ the other tasks in $M - N$) must dominate $\overline{X}$ if tasks in $M - N$ have strictly monotonic preferences. This contradicts the assumption that $\overline{X}$ is APO. Therefore, we have $Y_0 = K$—which implies that X is NI.

[Only if part of (ii)]. Let $X = (X_1, X_2, \cdots, X_n) \in \mathcal{X}_N$ be WAPO at $\succsim_N$. We prove that (a) X is PO and (b) X is WNI.

Proof of (a): Suppose not. Then, there exists $Y = (Y_1, \cdots, Y_n) \in \mathcal{X}_N$ that dominates X. Suppose $\succsim_{M-N}$ is strictly anti-monotonic. Since X is WAPO, there exists a PO allocation $\overline{X} = (X_1, \cdots, X_m) \in \mathcal{X}_M$. Clearly, $X_{n+1} = \cdots = X_m = \phi$ in order for $\overline{X}$ to be PO. But then, $\overline{Y} := (Y_1, \cdots, Y_n, \phi, \cdots, \phi)$ dominates $\overline{X}$—a contradiction.

Proof of (b): Suppose not. Then, there exists $Y = (Y_1, \cdots, Y_n) \in \mathcal{X}_N$ that weakly dominates X and

$$\neg (Y_1 + \cdots + Y_n \supseteq X_1 + \cdots + X_n). \tag{2}$$

Since X is PO by (a), it must be $X_i \sim_i Y_i$ for all $i \in N$. Now, assume that each $\succsim_{n+1}, \cdots, \succsim_m$ ranks $L := K - (Y_1 + \cdots + Y_n)$ on the first, ϕ on the second, and other coalitions lower than them. At this profile, one can verify that no matter how we select $X_{n+1}, \cdots, X_m$, $\overline{X} = (X_1, \cdots, X_m)$ is dominated by $\overline{Y} := (Y_1, \cdots, Y_n, L, \phi, \cdots, \phi)$ (because from (1) and (1), we can derive that $X_{n+1}, \cdots, X_m \neq L$). This contradicts with X being WAPO.

[If part of (ii)]. Suppose that $X = (X_1, \cdots, X_n)$ is PO, and WNI—and yet not WAPO—at $\succsim_N$. Let $\succsim_{M-N} \in \mathcal{D}_{M-N}$ and $\succsim = (\succsim_N, \succsim_{M-N})$. The negation of WAPO implies that

$$\text{for all } X_{n+1}, \cdots, X_m \text{ such that } \overline{X} := (X_1, \cdots, X_m) \in \mathcal{X}_M,$$
$$\text{there exists } Y = (Y_1, \cdots, Y_m) \in \mathcal{X}_M \text{ that dominates } \overline{X} \text{ at } \succsim. \tag{3}$$

Since X is PO and WNI, $Y_1 + \cdots + Y_n \supseteq X_1 + \cdots + X_n$ holds. So, $(Y_{n+1}, \cdots, Y_m)$ is an $(M - N)$-allocation with the set of robots $L := K - (X_1 + \cdots + X_n)$.

Note that this argument guarantees that for all $(M - N)$-allocations $(X_{n+1}, \cdots, X_m)$ (with L being the set of robots), there exists $(M - N)$-allocation $(Y_{n+1}, \cdots, Y_m)$ (with L being the set of robots) that dominates X. As we showed in Observation 1, this can never happen due to the finiteness of N, M.   ∎

Several comments are in order:

(a) Sufficiency and weak sufficiency are "global" properties in the sense that they refer to the allocation of the global set M. On the other hand, whether an N-allocation is PO or (W)NI should be considered "local" because they can be judged only within N. In this sense, Proposition 2 and Proposition 3 connect such "global" and "local" views.

(b) Proposition 2 characterizes sufficient local profiles with the existence of local allocations being PO and (W)NI; meanwhile, Proposition 3 is a parallel result on local allocations instead of local profiles. Although similar, PO and (W)NI are—in general—logically independent (see Proposition 5, Observation 2, and Observation 3 below).

(c) It is worth noting that the "only if" parts of Proposition 3 specify the possible local allocations that can be extended to global PO allocations. According to (i) of Proposition 2, APO allocation $\overline{X} \in \mathcal{X}_M$ assigns no robots to the tasks in $M - N$. In other words, APO allocation assigns all robots to well-known tasks (tasks in N) and leaves no robots standing by for future use (use for tasks in $M - N$). On the other hand, some robots can be unassigned under WAPO N-allocation.

(d) It is also worth noting that if a local profile $\succsim_N$ is sufficient, it remains sufficient after the matchmaker gets to know the preferences of the unknown tasks. However, this does not hold for weak sufficiency (see Proposition 4).

Proposition 4: Extending (weak) sufficient profiles

(i) If $\succsim_N \in \mathcal{D}_N$ is sufficient and $\succsim_{M-N} \in \mathcal{D}_{M-N}$, then $(\succsim_N, \succsim_{M-N})$ is also sufficient.

(ii) Even if $\succsim_N \in \mathcal{D}_N$ is weakly sufficient, and $\succsim_{M-N} \in \mathcal{D}_{M-N}$, $(\succsim_N, \succsim_{M-N})$ can fail to be weakly sufficient.

Proof of Proposition 4

(i) Suppose $\overline{X} = (X_1, \cdots, X_m) \in \mathcal{X}_M$ is APO at $\succsim_N$. Then, $\overline{X}$ is PO at $(\succsim_N, \succsim_{M-N})$. If $\overline{Y} = (Y_1, \cdots, Y_m) \in \mathcal{X}_M$ weakly dominates $\overline{X}$, $Y = (Y_1, \cdots, Y_n)$ must weakly dominate $X = (X_1, \cdots, X_n)$. Since X is NI by (i) of Proposition 3, it follows that $Y_1 + \cdots + Y_n = K$. Therefore, $\overline{X}$ is NI. By (i) of Proposition 2, $(\succsim_N, \succsim_{M-N})$ is also sufficient.

(ii) Let $a, b \in K$ be distinct robots. Assume that each task in $M \setminus \{m\}$ has strictly anti-monotonic preferences, and that $\{a\} \sim_m \{b\} \succ_m \phi \succ_m \cdots$. Then, $\succsim_N$ is weakly sufficient because $X = (\phi, \cdots, \phi) \in \mathcal{X}_N$ is PO and WNI—but $(\succsim_N, \succsim_{M-N})$ is not. Under $\succsim$, a PO M-allocation must assign ϕ to every task except m, and assign $\{a\}$ or $\{b\}$ to task m. In either case, such an M-allocation is not WNI. ∎

Proposition 5 (NI under PO)

Under the linear domain $\mathcal{D} = \mathcal{L}(\mathcal{K})^n$, if an N-allocation X is PO, then it is NI if—and only if— $X_1 + X_2 + \cdots + X_n = K$.

Proof of Proposition 5.

The "only if" part is straightforward. For the "if" part, assume that $X \in \mathcal{X}_N$ is PO and $X_1 + \cdots + X_n = K$. If $Y \in \mathcal{X}_N$ weakly dominates X, $Y_i \sim_i X_i$ for all $i \in N$ (otherwise, Y strictly dominates X). Since $\succsim_i$ is a linear order, $Y_i = X_i$ for all $i \in N$. ∎

Note that this does not hold when the domain is general. Let $N = \{1\}$, $K = \{a\}$, and $\{a\} \sim_1 \phi$. Then, $X = (\{a\})$ is clearly PO but not NI.

Observation 2 (Independence of PO and NI)

Under either general domain $\mathcal{D} = \mathcal{R}(\mathcal{K})$ or linear domain $\mathcal{D} = \mathcal{L}(\mathcal{K})$, PO and NI—as properties of allocations—are logically independent. The following examples prove this.

(a) N-allocation that is PO but not NI: If every task in N has strictly anti-monotonic preferences, $X = (\phi, \cdots, \phi) \in \mathcal{X}_N$ is surely PO but not NI.

(b) N-allocation that is NI but not PO: Let $N = \{1, 2, \cdots\}$ and $K = \{a, b, \cdots\}$ with

$$\{b\} \succ_1 \{a\} \succ_1 \cdots, \{a\} \succ_2 \{b\} \succ_2 \cdots, \text{ and } \phi \succ_j \cdots (j \neq 1, 2).$$

Then, $(X_1, X_2, \cdots) = (\{a\}, \{b\}, \cdots)$ is NI and not PO (dominated by $(\{b\}, \{a\}, \cdots)$).

Observation 3 (Relationship between PO and WNI)

(i) Under either the general domain or linear domain, WNI does not necessarily imply PO. This is already shown in (b) of Observation 2.

(ii) Under the general domain, PO and WNI are logically independent. The following example (as well as (i)) prove this fact: Let $a \in K$. If $\{a\} \sim_i \phi \succ_i \cdots$ for all $i \in N$, then $X = (\{a\}, \phi, \cdots)$ is PO but not WNI ($\because Y := (\phi, \phi, \cdots)$ weakly dominates X).

(iii) Under the linear domain, PO implies WNI (but not vice versa, by (i)). To prove this, assume that $X = (X_1, \cdots, X_n) \in \mathcal{X}_N$ is a PO N-allocation. If $Y = (Y_1, \cdots, Y_n) \in \mathcal{X}_N$ weakly dominates X, it must be $X_i \sim_i Y_i$ for all $i \in N$ (otherwise, Y dominates X; and so X cannot be PO). This implies that $Y_i = X_i$ for all $i \in N$ under the linear domain. Therefore, X is WNI. ∎

5 Allocation Rule to Obtain (W)APO Allocations

In this section, we study how to find a (W)APO allocation. The set of all permutations of a finite set A is denoted by $\Sigma(A)$.

Definition 4 (Sequential choice rule and serial dictatorship)
An allocation rule $f : \mathcal{D}_N \to \mathcal{X}_N$ is called a *sequential choice rule (SCR)* if there exists $\sigma : \mathcal{D}_N \to \Sigma(A)$ such that for all $\succsim = (\succsim_1, \cdots, \succsim_n) \in \mathcal{D}_N, f(\succsim) = (X_1, \cdots, X_n)$ satisfies the following (for the ease of notation, we write $\sigma_{\succsim}$ instead of $\sigma(\succsim)$):
(i) $X_{\sigma_{\succsim}(1)}$ is one of the greatest elements among $\mathcal{K}$, with respect to $\succsim_{\sigma_{\succsim}(1)}$.
(ii) For $j = 2, 3, \cdots, n, X_{\sigma_{\succsim}(j)}$ is one of the greatest elements among

$$\left\{ C \mid C \subseteq K \backslash \left(X_{\sigma_{\succsim}(1)} + X_{\sigma_{\succsim}(2)} + \cdots + X_{\sigma_{\succsim}(j-1)} \right) \right\}$$

—with respect to $\succsim_{\sigma_{\succsim}(j)}$.
An allocation rule $f : \mathcal{D}_N \to \mathcal{X}_N$ is called a *serial dictatorship (SD)* if it is SCR with the constant function σ.

In words, σ is a permutation that represents the precedence order of the tasks. The allocation $f(\succsim)$ is determined as follows. Task of the first order, $\sigma_{\succsim}(1)$, selects one of its best coalitions of robots. For $j \geq 2$—the task of the jth order, $\sigma_{\succsim}(j)$, selects one of its best coalitions of robots from the remaining robots. The precedence order $\sigma_{\succsim}$ can vary in SCR; however, it is constant in SD. Under the linear domain, our definition is equivalent to the standard definitions of both SCR and SD (cf., [9]). However, our definition is slightly generalized so that it can also deal with general domains. Under general domains with a given σ, the SCR (SD) represents a class of allocation rules; this is because more than one of the coalitions can be the greatest elements.

Proposition 6 (Sequential choice rule cannot be NI)
No SCR is NI.

Proof of Proposition 6: Suppose that $N = \{1, 2, \cdots\}$ and $K = \{a, b, c, \cdots\}$. Let $\{a, b\} \succ_1$ $\{a\} \succ_1 \phi \succ_1 \cdots, \{a, b\} \succ_2 \{b, c\} \succ_2 \phi \succ_2 \cdots$, and $\phi \succ_j \cdots$ for $j \neq 1, 2$. Then, $(\{a\}, \{b, c\}, \phi, \cdots)$ is the unique NI allocation; however, no SCR can yield this allocation because—under any SCR—either task 1 or 2 selects $\{a, b\}$.   ■

Proposition 7 (Serial dictatorship)
(i) Under the general domain, there exists an SD that is neither PO nor WNI.
(ii) Under the linear domain, any SD is WNI.

Proof of Proposition 7:
   Let us assume $\sigma(i) = i$ for all $i \in N$ (this does not lose the essence of the proof).
(i) Let $N = \{1, 2, \cdots\}$ and $K = \{a, b, \cdots\}$. Suppose that $K \sim_1 \phi \succ_1 \cdots, \{a\} \succ_2$ $\phi \succ_2 \cdots$, and $\phi \succ_j \cdots$ for $j \neq 1, 2$. Then, $X = (K, \phi, \phi, \cdots)$ is obtained by a SD; however, it is dominated by $Y = (\phi, \{a\}, \phi, \cdots)$. Therefore, such SD is not PO. Furthermore, $X_1 + X_2 \cdots = K \supset Y_1 + Y_2 + \cdots$ implies that it is not WNI—either.

(ii) Suppose that $X = (X_1, \cdots, X_n) \in \mathcal{X}_N$ is obtained by SD, and $Y = (Y_1, \cdots, Y_n) \in \mathcal{X}_N$ weakly dominates X. By the definition of SD, $X_1 \succsim_1 Y_1$. Furthermore, $Y_1 \succsim_1 X_1$ also follows because Y weakly dominates X. Since $\succsim$ is supposed to be a linear order, this means that $X_1 = Y_1$. By the definition of SD, X_2 is—at least—as good as any element in $2^{K \setminus X_1}$. Since $X_1 = Y_1$, we can also say that $X_2 \succsim_2 Y_2$. Furthermore, $Y_2 \succsim_2 X_2$ also follows because Y weakly dominates X. Since $\succsim$ is supposed to be a linear order, this means $X_2 = Y_2$. In the same way, $X_j = Y_j$ ($j = 1, 2, \cdots, n$) holds. Therefore, $X = Y$. ■

6 Final Remarks

The present paper explores how to reach a global PO allocation of robots (indivisible objects) based on the local preference profile, showing some characterization results (Proposition 2 and Proposition 3). One possible concern is that not all profiles are sufficient or are even weakly sufficient (Proposition 4). Is this a serious problem? The answer depends on the expectations of the tasks' preferences. Suppose, for instance, that each linear order in $\mathcal{L}(\mathcal{K})$ is equally likely as the preferences of each task. In such a case, the probability that N includes—at least—one task with strictly monotonic preferences (recall that such a task makes $\succsim_N$ sufficient) goes to 1 as $|N|$ becomes large. On the other hand, if most of the tasks have non-monotonic preferences (e.g., because of the lack of work yard), then the local profile $\succsim_N$ can often fail to be sufficient. In this sense, the likelihood of finding a PO allocation largely depends on the probability models of the preference profiles. The design of attractive allocation rules under typical probability models can be an interesting topic for future discussion.

Acknowledgements. This work is supported by JST [Moonshot Research and Development], Grant Number [JPMJMS2032].

References

1. Chevaleyre, Y., Dunne, P.E., Endriss, U., Lang, J., Maudet, N., Rodríguez-Aguilar, J.: Multiagent resource allocation. Knowl. Eng. Rev. **20**, 143–149 (2005). https://doi.org/10.1017/S0269888905000470
2. Chevaleyre, Y., et al.: Issues in multiagent resource allocation. Inform **30**, 3–31 (2006)
3. Shapley, L., Scarf, H.: On cores and indivisibility. J. Math. Econ. **1**, 23–37 (1973). https://doi.org/10.7249/r1056-1
4. Darmann, A.: Group activity selection from ordinal preferences. In: Walsh, T. (ed.) ADT 2015. LNCS (LNAI), vol. 9346, pp. 35–51. Springer, Cham (2015). https://doi.org/10.1007/978-3-319-23114-3_3
5. Darmann, A., Elkind, E., Kurz, S., Lang, J., Schauer, J., Woeginger, G.: Group activity selection problem. In: International Workshop on Internet and Network Economics, pp. 156–169 (2012)
6. Gerkey, B.P., Matarić, M.J.: A formal analysis and taxonomy of task allocation in multi-robot systems. Int. J. Rob. Res. **23**, 939–954 (2004). https://doi.org/10.1177/0278364904045564
7. Brams, S.J., Kilgour, D.M., Klamler, C.: How to divide things fairly. Math. Mag. **88**, 338–348 (2015). https://doi.org/10.4169/math.mag.88.5.338

8. Brams, S.J., Kilgour, D.M., Klamler, C.: Maximin envy-free division of indivisible items. Gr. Decis. Negot. **26**, 115–131 (2017). https://doi.org/10.1007/s10726-016-9502-x
9. Pápai, S.: Strategyproof and nonbossy multiple assignments. J. Public Econ. Theory. **3**, 257–271 (2001)
10. Ehlers, L., Klaus, B.: Coalitional strategy-proof and resource-monotonic solutions for multiple assignment problems. Soc. Choice Welfare. **21**, 265–280 (2003). https://doi.org/10.1007/s00355-003-0259-1
11. Hatfield, J.W.: Strategy-proof, efficient, and nonbossy quota allocations. Soc. Choice Welfare. **33**, 505–515 (2009). https://doi.org/10.1007/s00355-009-0376-6
12. Manea, M.: Serial dictatorship and Pareto optimality. Games Econ. Behav. **61**, 316–330 (2007). https://doi.org/10.1016/j.geb.2007.01.003
13. Rastegari, B., Condon, A., Immorlica, N., Leyton-Brown, K.: Two-sided matching with partial information. In: Proceedings of the AAAI Conference on Artificial Intelligence, vol. 28, no. 1, pp. 733–750 (2014). https://doi.org/10.1145/2482540.2482607
14. Drummond, J., Boutilier, C.: Preference elicitation and interview minimization in stable matchings. In: Proceedings of the AAAI Conference on Artificial Intelligence, vol. 28, no.1, pp. 645–653 (2014)
15. Aziz, H., Biró, P., Gaspers, S., de Haan, R., Mattei, N., Rastegari, B.: Stable matching with uncertain linear preferences. In: International Symposium on Algorithmic Game Theory, pp. 195–206 (2016). https://doi.org/10.1007/s00453-019-00650-0
16. Puppe, C.: An axiomatic approach to "preference for freedom of choice." J. Econ. Theory. **68**, 174–199 (1996). https://doi.org/10.1006/jeth.1996.0009
17. Barberà, S., Bossert, W., Pattanaik, P.K.: Ranking sets of objects. Handb. Util. Theor., 893–977 (2004). https://doi.org/10.1007/978-1-4020-7964-1_4

Neuroscience Behavioral Studies for Modulation of the FITradeoff Method

Lucia Reis Peixoto Roselli[✉] ⓘ and Adiel Teixeira de Almeida ⓘ

Center for Decision Systems and Information Development (CDSID),
Universidade Federal de Pernambuco, Recife, PE, Brazil
{lrpr,almeida}@cdsid.org.br

Abstract. It has been claimed in the literature that decision-making methods have not been modulated (transformed) by results obtained in behavioral studies as much as has been expected and that further modulation would be an important advancement in decision-making. This paper summarizes the modulation provided by the Flexible and Interactive Tradeoff (FITradeoff) method from behavioral studies performed using neuroscience tools. Modulations of the FITradeoff method have been conducted in two ways: modulations in the preference modelling process and modulations in the FITradeoff Decision Support System (DSS). For modulation in FITradeoff preference modeling, several recommendations were provided to support analysts during their advising process with decision-makers. For modulation in the FITradeoff DSS, several improvements were implemented in the design of the DSS. The modulation of the FITradeoff method was supported by neuroscience experiments. These experiments investigated decision-makers' (DMs) behavior when they interacted with a holistic evaluation and elicitation by decomposition in the FITradeoff method. The modulation of the FITradeoff method promoted the inclusion of some features through the combination of the two paradigms of preference modeling, completely transforming the decision-making process, and its DSS.

Keywords: Modulation · FITradeoff method · Preference modeling · Decision support system · Behavioral studies

1 Introduction

Korhonen & Wallenius [1] suggested that decision-making methods should be modulated (transformed) by the consideration of behavioral aspects involved in decision processes. Discussion on behavior has emerged as an important aspect of advanced research in decision-making [2, 3].

An alternative to modulating decision-making methods and procedures has been proposed using a Neuroscience approach to investigate decision-makers' (DMs) behavior in several decision-making experiments [4, 5]. In an editorial of a special issue related to the use of Neuroscience in Decision-Making, it has been stated that the kind of study presented in this paper is "… an interesting direction for the development of effective decision support systems …" [6].

D. C. Morais and L. Fang (Eds.): GDN 2022, LNBIP 454, pp. 44–58, 2022.
https://doi.org/10.1007/978-3-031-07996-2_4

Different areas of knowledge can be advanced using neuroscience tools. For decisions in games, the Neuroeconomics have been developed [7, 8], for consumer and marketing decisions, studies about Consumer Neuroscience and Neuromarketing have been performed [9, 10]. Regarding information science, studies on NeuroIS have been conducted [11, 12]. This shows how modulation of methods can be conducted, particularly for the Flexible and Interactive Tradeoff (FITradeoff) method [13, 14], supported by neuroscience tools.

Considering Multi-Criteria Group Decision and Negotiation (GDN), Group Decision (GD) and Negotiation (N) are approaches that present many similarities which cannot be completely dissociated [15]. However, there are some differences between GD and N regarding how behavioral studies can be conducted, specifically considering preference modeling and interaction processes with negotiation procedures and tools [16–22].

It is worth to mention that in the scope of Multi-Criteria Group Decision Making/Aiding (MCGDM/A), behavioral consideration effectively investigates how each Decision-Maker (DM) expresses his/her preferences in order to integrate the group preferences subsequently, since MCGDM/A encompass several preferences to generate the group preferences. Thus, the modulation was conducted considering individual preferences expressed by each DM as it was assumed that more confident information could be collected for the preference modeling process. That is, behavioral consideration effectively investigates how DMs express their preferences rather than those generated by the group [16]. One can argue that behavioral studies in GD could not be applied to the whole group of DMs at once without affecting the confidence of the preference modeling process. It is already hard for a unique DM to interpret the process, making reflections for supplying information confidently. |One can image how harder it would be to make it collectively, mixing the own interpretation process at the same time of hearing preference statements by the other DMs.

Hence, the motivations in developing this paper have been to synthesis behavioral studies concerning GD, especially those which investigate the FITradeoff method [13, 14]. Therefore, this paper summarizes the studies performed using neuroscience tools to modulate the FITradeoff method in two segments:

- Improving the preference modeling process
- Improving the FITradeoff Decision Support System (DSS)

The former concerns the elaboration of a set of recommendations that were generated from investigations of DMs during the FITradeoff decision-making process. These outcomes can be used by analysts to support and advise DMs. The latter concerns improvements in the design of the FITradeoff DSS, specifically, the inclusion of messages, buttons, graphs, and tools to improve DMs' experience using the FITradeoff software. The FITradeoff DSS is web-based and available free of charge at www.fitrad eoff.org.

In this context, the contribution of this paper is to discusses important behavioral studies concerning the FITradeoff method to provide several recommendations for Decision-Makers and Analysts. Hence, based on this paper recommendations can be followed by DMS to support the decision process with the FITradeoff method. Moreover, this paper discusses about new features which were implemented in the FITradeoff DSS to promote a better journey for DMs [14]. The innovation of this study is to compose studies concerning the use of neuroscience tools to investigate decision-making method [23, 24, 28], being important to advance studies in Decision Neuroscience area.

The paper is structured as follows: Sect. 2 briefly describes the FITradeoff method and Sect. 3 summarizes the experiments performed to investigate it. Section 4 describes the main results obtained from these experiments. Thus, Sects. 5 and 6 present several recommendations to modulate the FITradeoff method in two aspects: modulation in preference modeling (Sect. 5) and modulation in the FITradeoff DSS (Sect. 6). Finally, Sect. 7 remarks the conclusion and future research.

2 The FITradeoff Method

The FITradeoff method [13, 14] is a method in the Multi-Attribute Value Theory (MAVT) context [29]. This method uses additive aggregation to elicit scaling constants. It is based on the Tradeoff Procedure [29] but incorporates partial information's about DMs preferences. In this context, using the FITradeoff method, DMs do not need to express indifference relations, as in Traditional Tradeoff procedure. Instead of that, they are required to express only strict preferences.

When DMs use the FITradeoff method, they can express preferences in two ways: during the elicitation by decomposition or during the holistic evaluation. In other words, the method combines the two perspectives of preference modeling. Hence, during the decision-making process, DMs can compare consequences in the elicitation by decomposition or they can compare alternatives during the holistic evaluation. The FITradeoff method is considered flexible since DMs can express preferences using both paradigms for preference modeling and can alternate between them, using whichever is more suitable for their cognitive style.

Each preference expressed by DMs in either decomposition or holistic evaluation is included in a Linear Programing Problem (LPP). Therefore, during each interaction with DMs, the LPP model is run, and the scaling constant space is updated. Hence, using the FITradeoff, the exact values of scale constant are not obtained, instead of that a space of available values is obtained at the final.

Also, FITradeoff is consider interactive since DMs participate in the role process expressing their preferences and can evaluate partial results after each interaction. In addition, they can interrupt the process when they judge partial results as sufficient.

The FITradeoff method was developed to support choice problems [13]. However, it has been extended to ranking [30], sorting [31], and portfolio problematic [32]. This method has been implemented in Decision Support System (DSS), which is available at www.fitradeoff.org. Hence, supported by the FITradeoff method several MCDM/A situations were solved. For supplying selection, the FITradeoff method has been applied to select the best supplier for a food industry [33], and in a Wholesaler and Retailer Company [34]. The method has also been used to select the best location for a healthy care unit [35] and a police station [36]. Moreover, in the context of health care, Camilo et al. (2020) [37] used the method to support the triage problem, and [38] to understand patient adherence behavior. For industry context, the method was applied to rank workstations in a footwear industry [39], to support schedule decisions [40], to support the (WCM) application [41], in select agricultural technology packages [42], to investigate business process management (BPM) [43], and to support the assessment of information systems [44]. The method has been also applied to support problems in energy and water context [45–48].

Moreover, behavioral studies have been developed to investigate the preference modeling process in the FITradeoff method, specifically in decomposition [49–51] and holistic evaluation [52–59]. The following section describes the behavioral experiments constructed to modulate the FITradeoff method.

3 Behavioral Experiments

All the experiments developed to modulate the FITradeoff method have been conducted at the NeuroScience for Information and Decision (NSID) laboratory at the Federal University of Pernambuco (UFPE), Recife, Brazil. Two types of experiments have been developed. One type has been constructed to investigate the preference modeling with elicitation by decomposition. The other has been developed to investigate the holistic evaluation phase in the FITradeoff method. These experiments were supported by two neuroscience tools: the 14-channel electroencephalograph (EEG) by Emotiv and the X120 eye-tracking device by Tobii Studio.

Since 2017 these experiments have been performed. The samples were composed of graduate and postgraduate management engineering students. The participants attended multi-criteria decision-making classes at UFPE, in which the experiment was an extra-evaluation. These students present similar knowledge about MCDM/A approach, and none had phycological problems that compromised the execution of these experiments. Moreover, a structured protocol was developed, and it was followed to perform the experiments with each one of the participants. This protocol was approved by the Ethical Committee in Research of the Federal University of Pernambuco with CAAE with number 31065820.5.0000.5208.

Three experiments were conducted to investigate DMs' behavior during the elicitation by decomposition—the first one in 2018 and the final one in 2021 [49–51]. In these experiments, the participants proposed their own multi-criteria decision problems in multi-criteria classes. In other words, they do not used a predefined case to solve. After that, the problems were reviewed to confirm that the criteria and alternatives were well-defined. Since, when the decision-matrix is concluded, the participate is ready to schedule their visit at the NSID lab. Hence, when they arrived at NSID lab, one per time, they receive the protocol recommendations. After that, they performed the experiment, in which they had to solve their decision problems using the decomposition process at FITradeoff DSS. The experiments used a similar version of the DSS available at www.fitradeoff.org., one constructed specifically for experimentation. Figure 1a illustrates an experiment conducted to investigate elicitation by decomposition.

Five experiments were conducted to investigate the holistic evaluation [52–59]. The first one was performed in 2017; it was a prelaminar experiment that only involved selecting the best alternative [57, 58]. These experiments used the graphics presented in the FITradeoff DSS (bar, spider, and bubble graphs). Hence, in these experiments, participants evaluated graphical and tabular visualizations and express dominance relations between the alternatives presented in them. In other words, for each graphical or tabular visualization, the participants should evaluate the performance of the alternatives considering the compensations between them, following the MAVT concepts [29], i.e., answering which is the best alternative or the worst alternative. The best and the worst alternative were previously computed following the MAVT concepts. For instance, for a bar graph, the participants should evaluate the heights of the bar and define which is the best one or the worst one. Figure 1b illustrates the experiments conducted to investigate the holistic evaluation.

These visualizations were constructed to represent MCDM/A problems which do not have a specific context, where the alternatives were identified by letters (A and B), and the criteria were identified by numbers (e.g. "Crit1"). The criteria weights were equal or different. The weights are computed by an arithmetic progression. For visualizations which had different values for the weights, the Crit 1 present the highest value for the weight and the Crit 5 the lowest value, from the left to the right,. In the experiments, the visualizations were positioned in a random sequence, and it was showed one per time. The same visualizations were used in an experiment. To see more details about visualizations creation, please see [57, 58].

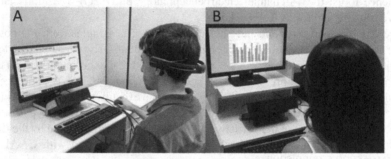

Fig. 1. a: Experiment to investigate elicitation by decomposition (Silva et al., 2021 [9]). b: Experiment to investigate holistic evaluation (Roselli et al., 2019 [57]).

It should be noted that an online experiment was also conducted during the COVID-19 pandemic in 2020. This experiment did not use the neuroscience tool but collected behavioral variables concerning holistic evaluation [59].

After the experiments have been applied, especially those that use neuroscience tools, physiological variables that are not controlled by participants, such as brain activity and eye movement, were captured. Based on these data, analyses were performed, and some interesting results were obtained, which are described in the next section.

4   Results obtained from Behavioral Experiments

From the holistic evaluation experiments, an important result generated was the Success Based Decision Rule (SBDR), in detail in the original paper [53]. This rule indicates probabilities of success concerning the selection of the best alternative using graphical and tabular visualizations. This rule is based on the hit rate (HR) variable, applying the Bernoulli model, which fits well in the process. Thus, the recommendations are derived by combining the probability of success (π) and the standard deviation (σ), based on the Bernoulli distribution.

Concerning the SBDR, it was also observed that DMs generally performed better when selecting the best alternative for visualizations that had similar values for the criteria weights [57, 58]. It is worth to mention, that the FITradeoff method is used to elicit the scaling constants (weights). However, using this method the exact values are not obtained, instead of that an available weight space is obtained at the final of the decision process. Hence, the weights distribution can be observed in the weights space. Moreover, based on the heuristic proposed in the original FITradeoff paper [13], the FITradeoff detects the distribution form for the weights after the first answer provided by Decision-Makers to the first question in the elicitation by decomposition (illustrated in Fig. 1a). Thus, the weights can present skewed or similar distribution.

Additionally, regarding the use of visualizations in the FITradeoff DSS, the neuroscience experiments suggested that tables should be included in the FITradeoff DSS, since participants presents higher hit when they used tables to express dominance relations between alternatives. In the same perspective, based on the Hit-Rate, the Spider graphs were not generally recommended to be used to express preferences between alternatives. In FITradeoff DSS, the table is the decision-matrix [29]. However, it only showed the alternatives selected by DMs in that moment of the decision process.

Another important result obtained using eye-tracking was that DMs focused more on criteria on the left side of the screen, which presented higher weights values. The weights values decreased from left to right in graphical and tabular visualizations (Fig. 2). Otherwise stated, DMs fixed firstly in the left side of the screen, moving their eye to the right side of the screen, following the criteria weights distribution [57, 58].

Moreover, with the use of the EEG, an important tool was created - the Alpha-Theta Diagram. This diagram is based on frontal Theta (4–8 Hz) and parietal Alpha (8–13 Hz) activities. Thus, based on this diagram, it is possible to classify DMs' behavior during decision-making concerning cognitive effort and engagement [52].

For experiments exploring the decomposition process [49], it was observed that participants spent more time performing elicitation by decomposition (Step 3) than on the ordering of criteria weights and consequence space examination (Steps 1 and 2) [29]. It was also observed that participants' pupil size increased during the steps' evolution. As an increase in pupil diameter suggests an increase in cognitive effort [60], this result suggests that higher cognitive effort was demanded to perform the decomposition step.

In these experiments, MCDM/A problems were divided into three groups: problems with quantitative criteria, problems with qualitative criteria, and problems with a combination of quantitative and qualitative criteria. A specific category was defined if at least 75% of the criteria were quantitative or qualitative, if not, the problem presented combined criteria. Hence, larger pupil size was observed in problems with combined criteria. Moreover, from brain activities in theta, alpha, and beta bands, it was suggested that combined problems demanded more cognitive effort from DMs [61]. Therefore, based on the results obtained from the behavioral studies described here, several recommendations are provided to modulate the FITradeoff method in two aspects: modulation in preference modeling (discussed in the next section) and modulation in the FITradeoff DSS (discussed in Sect. 6).

5 Modulation in the Preference Modeling Process

For modulation in the preference modeling of the FITradeoff method, several recommendations are elaborated upon from the behavioral studies to support analysts in the advisory process with DMs. In this section, the results presented above are discussed concerning recommendations for analysts.

In this context, concerning to the Success Based Decision Rule (SBDR), this rule was already included in the FITradeoff DSS on the analyst page. The analyst tab in FITradeoff DSS can be only accessed by analysts, who has methodological knowledge about the FITradeoff method. However, DMs can also use the DSS without an analyst.

Thus, during the decision-making process, when DMs evaluated graphical and tabular visualizations for holistic evaluation, the analyst could access the SBDR to consult the probability of success regarding these visualizations, all these probabilities are presented in [53]. Hence, it a visualization presents a lower probability of success, the analysts can recommend DMs to not use this visualization to express a preference relation, being recommended to test other type of visualization or return to the elicitation by decomposition to continue the preference modeling process. Therefore, based on this rule, the analyst can advise DMs on whether to use visualizations to express a preference relation between alternatives during the holistic evaluation.

Another recommendation for analysts is concerning problems with had similar values for the criteria weights. Since it was observed that DMs generally performed better using visualizations in this kind of problem, analysts should advise DMs to express more preferences in the holistic evaluation than in problems with "skewed" values for the weights [57, 58]. Thus, when DMs is faced to problems with "skewed" values for the weights, a suggestion for analysts is to advising DMs to spend more time and pay attention in analyzing visualizations before expressing preferences during holistic evaluation.

Also, for spider graphs, analysts should advise DMs to take care in using this kind of visualization to express preferences and perhaps use bar graphs instead, for instance.

Concerning the Alpha-Theta Diagram [52], the analyst can use this tool to classify DMs' patterns of behavior, considering cognitive effort and engagement. The Alpha-Theta Diagram is not specific to the FITradeoff context but can be applied to several MCDM/A methods in the context of MAVT to support the advising process performed by the analyst.

Particularly, for the FITradeoff method, the results obtained using the Alpha-Theta Diagram indicated that DMs evaluated graphical visualizations with high cognitive effort and/or high engagement during holistic evaluation. Hence, this result reinforces the use of graphical visualizations in holistic evaluations and the combination of the two paradigms of preference modeling in FITradeoff decision processes. Thus, analysts can advise DMs to combine these two paradigms, altering how preferences are expressed during the FITradeoff decision-making process since it provides flexibility and possibly reduces the time spent obtaining a solution.

Moreover, considering the flexibility presented in the FITradeoff decision-making process by the combination of these two perspectives of preference modeling, another important result to modulate the FITradeoff decision-making process concerns the inclusion of the elimination process in the holistic evaluation. DMs can select the alternative which dominates the other or can eliminate a dominated alternative for the group. This flexibility is useful for choice problems in which DMs can condense the group of Potentially Optimal Alternatives (POA) by performing a holistic evaluation. Therefore, analysts should observe which decision-making processes (selection or elimination of alternatives) are the most suitable for DMs' cognitive styles to better advise them when evaluating visualizations during the FITradeoff process. It is not a trivial process, depends on analyst experience to conduct a decision process. But analyst should observe how DMs feel when they are asked to select versus when they are asked to eliminate an alternative.

Finally, from the specific results obtained in the experiments which investigated the decomposition process, it was observed that more cognitive effort and time were demanded in the decomposition step than in the previous step. Moreover, problems with both quantitative and qualitative criteria also demanded more time and cognitive effort to proceed in the FITradeoff method. Thus, analysts should advise DMs to conduct elicitation by decomposition with more attention for problems with both qualitative and quantitative criteria, which are very common in the MCDM/A context.

Hence, a modulation for preference modeling in the FITradeoff method was done by several recommendations provided by the analyst to support DMs during the FITradeoff decision-making process to obtain solutions per their true preferences. The following section discusses the improvements implemented to modulate the FITradeoff DSS.

6 Modulation in the DSS

The modulation of the FITradeoff DSS indicates the improvements implemented in the design of the FITradeoff software, specifically, the inclusion of messages, buttons, graphs, and tools to improve DMs' experience when using the FITradeoff DSS.

The first improvement in the design of the FITradeoff DSS was the inclusion of analyst access (login and password) and several recommendations on the analyst page. Therefore, during the FITradeoff decision-making process, analysts could consult the analyst page to review the recommendations provided to support DMs.

The SBDRL [53] was included on the analyst page. Hence, during the holistic evaluation, the analyst could click on the analyst page and seek advice from the use of a specific visualization to perform the holistic evaluation. From that page, the analyst could obtain recommendations for all types of visualizations presented in the FITradeoff DSS (bar graphs, spider graphs, bubble graphs, and tables) from the combination of alternatives and criteria which had been tested in the neuroscience experiments [52–59]. It should be noted that the recommendation was seen for one visualization at a time. More recommendations were uploaded to the analyst page as an output of holistic evaluation experiments.

Another important conclusion is that DMs moved their eyes from the left to the right side of the screen following the distribution of the weights. Since in experiments, the criterion which had the highest value for the weights is in the left side of the screen and the criterion which had the lowest value was in the right side. Thus, this confirms that the weights were correctly distributed in the visualizations presented on the FITradeoff DSS, from the left to the right. Figure 2 illustrates a heat map obtained using eye-tracking for a bar graph with three alternatives and four criteria; the red points indicate the areas where DMs focused their eyes more.

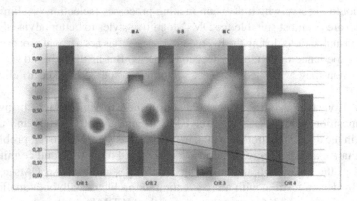

Fig. 2. Heat map to illustrate eye-movements

As discussed in the previous section, some of the recommendations made by the analyst utilized information regarding the distribution of weights [57, 58]. Hence, in the FITradeoff DSS, a note was included below the visualizations to indicate the distribution of criteria weights Buttons were not used to avoid distracting DMs from the holistic evaluation screen.

Four visualizations are presented in the DSS for choice and ranking problematic. In the begging, only graphical visualizations were used for choice problematic. However, a modulation was conducted to ranking problematic, in which the visualizations were included to support the holistic evaluations. Moreover, tables are considered for ranking and choice problematic.

In the case of ranking problematic, the holistic evaluation has been performed to supported DMs to comparing alternatives in the same ranking position, but still incomparable at that stage of the decision-making process. Otherwise stated, as FITradeoff uses partial information, alternatives can sometimes be incomparable. Hence, the DMs can select the pair of incomparable alternatives that they want to express a dominance relation Therefore, using the visualizations, DMs could compare pairs of incomparable alternatives and, if they desired, define a relationship between them.

Another important improvement developed in the DSS concerning ranking was the indication in the Hasse Diagram of the dominance relation defined during the holistic evaluation. Thus, the Hasse Diagram highlighted in red shows preference relations defined in the holistic evaluation and highlighted in black shows preference relations defined in the elicitation by decomposition, as illustrated in Fig. 3.

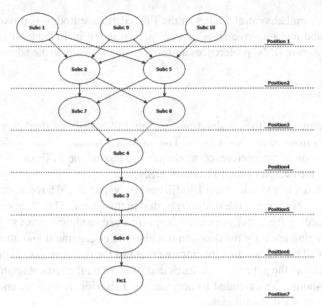

Fig. 3. Hasse Diagram with red arrow to show the preference expressed in the holistic evaluation (Barla et al. [62])

Finally, another important modulation in the FITradeoff DSS concerns the inclusion of an additional inequality in the LPP model of the FITradeoff method [14]. This inequality refers to the preference relations that can be defined during the holistic evaluation, as illustrated in Eqs. (1), in which indicates that the alternative Aj dominates the Alternative Ak.

$$\sum_{j=1}^{m} k_j v_j(A_i) > \sum_{j=1}^{m} k_j v_j(A_k) \tag{1}$$

When the FITradeoff is developed [13], preferences expressed during the holistic evaluation was used only to finalize the decision process. This modulation, provided from the behavioral studies, transforms the FITradeoff decision-making process and provides a new feature for this method—combining the two paradigms of preference modeling [14].

Now, in the FITradeoff DSS, the preferences expressed during holistic evaluation are also included in the LPP model to provide additional information for the elicitation process. Moreover, the DSS permits DMs to select the dominant alternative or that which is dominated by the other alternatives.

The design of the FITradeoff DSS was modulated using this combination of paradigms. In the recent version of the DSS, DMs can alternate between these paradigms from the beginning of the decision-making process. The elicitation by decomposition screen has also been modulated to highlight this combination, allowing the holistic evaluation to be conducted clearly and flexibly.

Therefore, from behavioral studies in the FITradeoff method, it is possible to modulate this method and improve DMs' experience of using it. FITradeoff is now more adequate at capturing DMs' preferences and supporting analysts in the advisory process.

7   Conclusion

This paper summarized the modulations to the FITradeoff method from the main results obtained from behavioral studies. The modulations have been conducted in two ways: modulations to the preference modeling process of the FITradeoff method and modulations to the design of the FITradeoff DSS.

In their editorial of a special issue Tikidji-Hamburyan et al. [6] have made an interesting remark concerning one of the studies reported in this paper [51]. They stated that it is interesting regarding the improvement of the preference modeling process "... because it allows optimizing not only the decision but also the engagement and attention of the decision maker" This is regarding the improvement of the preference modeling process. They also stated that the paper "... indicates that modern mathematical approaches may, and probably should, be extended to account for behavior as well as emotional and cognitive states of the decision maker.".

For modulations in FITradeoff preference modeling, the focus was on the analyst's advisory process. Thus, several recommendations have been suggested to support analysts in their interactions with DMs. For modulations to the design of the FITradeoff DSS, the focus was on the inclusion of many tools to support the decision process, such as notes, buttons, visualizations, and arrows.

The main modulation concerns the combination of the two paradigms of preference modeling in the FITradeoff, which completely transforms the decision-making process. In previous versions of the FITradeoff method [13], the preference expressed during the holistic evaluation could only be used to finalize the decision-making process. However, now, the preferences expressed in either decomposition or holistic evaluation have been included in the LPP model. Additionally, DMs can vary how they express preferences, using those that they deem most appropriate for their cognitive style. This combination promotes a redesign of DSSs for choice and ranking problematic and how analysts should conduct the process with DMs.

Therefore, behavioral studies performed with the support of neuroscience tools have an important role regarding the modulation of this method. Moreover, these studies represent innovation in the MCDM/A area [1, 2], which until now do not present many studies in this theme [23].

For the modulation of the FITradeoff method, the method has been improved to provide a better experience for DMs interacting with it. Now, the FITradeoff is in line with DMs' behavior regarding how they express preferences. Additionally, the method provides several recommendations to analysts, supporting this important agent in the decision-making field. The expectation is that, by using the FITradeoff method, DMs can obtain the solutions which are per their true preferences in an accurate, flexible, and interactive manner.

For future research, several experiments must be conducted to continue the modulation of the FITradeoff method. For instance, different participants should be interviewed to collect their preferences for different problems, including not only management engineering students. The experiments should test the combination of these two paradigms Several recommendations have already been included in the web version of the FITradeoff method.

Acknowledgment. This work had partial support from the Brazilian Research Council (CNPq) [grant 308531/2015-9;312695/2020-9] and the Foundation of Support in Science and Technology of the State of Pernambuco (FACEPE) [APQ-0484-3.08/17].

References

1. Korhonen, P., Wallenius, J.: Behavioral issues in MCDM: neglected research questions. In: Multicriteria Analysis, pp. 412–422. Springer, Heidelberg (1997). https://doi.org/10.1007/978-3-642-60667-0_39
2. Wallenius, J., Dyer, J.S., Fishburn, P.C., Steuer, R.E., Zionts, S., Deb, K.: Multiple criteria decision making, multiattribute utility theory: recent accomplishments and what lies ahead. Manage. Sci. **54**(7), 1336–1349 (2008)
3. Wallenius, H., Wallenius, J.: Implications of world mega trends for MCDM research. In: Ben Amor, S., de Almeida, A., de Miranda, J., Aktas, E. (eds.) Advanced Studies in Multi-Criteria Decision Making. Series in Operations Research, 1st ed., pp. 1–10. Chapman and Hall/CRC, New York (2020)
4. Zhao, Y., Zhao, X., Wang, L., Chen, Y., Zhang, X.: Does elicitation method matter? Behavioral and neuroimaging evidence from capacity allocation game. Prod. Oper. Manag. **25**(5), 919–934 (2016)
5. Smith, D.V., Huettel, S.: Decision neuroscience: neuroeconomics. Wiley Interdisc. Rev. Cogn. Sci. **1**(6), 854–871 (2010)
6. Tikidji-Hamburyan, R.A., Kropat, E., Weber, G.-W.: Preface: operations research in neuroscience II. Ann. Oper. Res. **289**, 1–4 (2020)
7. Glimcher, P.W., Rustichini, A.: Neuroeconomics: the consilience of brain and decision. Science **5695**, 447–452 (2004)
8. Fehr, E., Camerer, C.F.: Social neuroeconomics: the neural circuitry of social preferences. Trends Cogn. Sci. **11**(10), 419–427 (2007)
9. Khushaba, R.N.: Consumer neuroscience: assessing the brain response to marketing stimuli using electroencephalogram (EEG) and eye tracking. Expert Syst. Appl. **40**(9), 3803–3812 (2013)
10. Morin, C.: Neuromarketing: the new science of consumer behavior. Society **48**(2), 131–135 (2011)
11. Riedl, R., Davis, F.D., Hevner, A.R.: Towards a NeuroIS research methodology: intensifying the discussion on methods, tools, and measurement. J. Assoc. Inf. Syst. **15**(10) (2014)
12. Dimoka, A., Pavlou, P.A., Davis, F.D.: Neuro-IS: the potential of cognitive neuroscience for information systems research. In: 28th International Conference on Information Systems, Proceedings, Toulon, França, pp. 1–20 (2007)
13. de Almeida, A.T., Almeida, J.A., Costa, A.P.C.S., Almeida-Filho, A.T.: A new method for elicitation of criteria weights in additive models: flexible and interactive tradeoff. Eur. J. Oper. Res. **250**(1), 179–191 (2016)

14. de Almeida, A.T., Frej, E.A., Roselli, L.R.P.: Combining holistic and decomposition paradigms in preference modeling with the flexibility of FITradeoff. CEJOR **29**(1), 7–47 (2021). https://doi.org/10.1007/s10100-020-00728-z
15. Kilgour, D.M., Eden, C.: Handbook of Group Decision and Negotiation: Advances in Group Decision and Negotiation, vol. 4. Springer, Cham (2010). https://doi.org/10.1007/978-90-481-9097-3
16. de Almeida, A., Rosselli, L., Costa Morais, D., Costa, A.: Neuroscience tools for behavioural studies in group decision and negotiation. In: Kilgour, D.M., Eden, C. (eds.) Handbook of Group Decision and Negotiation, 1st edn., pp. 1–24. Springer International Publishing, Dordrecht, Netherlands (2020)
17. von Neumam, J., Morgenstern, O.: Theory of Games and Economic Behavioral, 3rd edn. Princeton University Press, Princeton (1953)
18. Raiffa, H.: The Art and Science of Negotiation: How to Resolve Conflicts and Get the Best Out of Bargaining. Harvard University Press, Cambridge (1982)
19. Schmid, A., Schoop, M.: Gamification of electronic negotiation training: effects on motivation, behaviour and learning. Group Decis. Negot., 1–33 (2022)
20. Roszkowska, E., Kersten, G.E., Wachowicz, T.: The impact of negotiators' motivation on the use of decision support tools in preparation for negotiations. Int. Trans. Oper. Res. (2021)
21. Engin, A., Vetschera, R.: Information representation in decision making: the impact of cognitive style and depletion effects. Decis. Support Syst. **103**, 94–103 (2017)
22. Vetschera, R.: Preference structures and negotiator behavior in electronic negotiations. Decis. Support Syst. **44**(1), 135–146 (2007)
23. Hunt, L.T., Dolan, R.J., Behrens, T.E.: Hierarchical competitions subserving multi-attribute choice. Nat. Neurosci. **17**(11), 1613–1622 (2014)
24. Nermend, K.: The implementation of cognitive neuroscience techniques for fatigue evaluation in participants of the decision-making process. In: Nermend, K., Łatuszyńska, M. (eds.) Neuroeconomic and Behavioral Aspects of Decision Making. SPBE, pp. 329–339. Springer, Cham (2017). https://doi.org/10.1007/978-3-319-62938-4_21
25. Özerol, G., Karasakal, E.: A parallel between regret theory and outranking methods for multicriteria decision making under imprecise information. Theor. Decis. **65**(1), 45–70 (2008)
26. Chuang, H., Lin, C., Chen, Y.: Exploring the triple reciprocity nature of organizational value cocreation behavior using multicriteria decision making analysis. Math. Problems Eng. **2015**, 1–15 (2015)
27. Trepel, C., Fox, C.R., Poldrack, R.A.: Prospect theory on the brain? Toward a cognitive neuroscience of decision under risk. Cogn. Brain Res. **23**(1), 34–50 (2005)
28. Barberis, N., Xiong, W.: What drives the disposition effect? An analysis of a long-standing preference-based explanation. J. Finan. **64**(2), 751–784 (2009)
29. Keeney, R.L., Raiffa, H.: Decisions with Multiple Objectives: Preferences, and Value Tradeoffs. Wiley, New York (1976)
30. Frej, E.A., de Almeida, A.T., Costa, A.P.C.S.: Using data visualization for ranking alternatives with partial information and interactive tradeoff elicitation. Oper. Res. Int. J. **19**(4), 909–931 (2019). https://doi.org/10.1007/s12351-018-00444-2
31. Kang, T.H.A., Frej, E.A., de Almeida, A.T.: Flexible and interactive tradeoff elicitation for multicriteria sorting problems. Asia Pac. J. Oper. Res. **37**, 2050020 (2020)
32. Frej, E.A., Ekel, P., de Almeida, A.T.: A benefit-to-cost ratio based approach for portfolio selection under multiple criteria with incomplete preference information. Inf. Sci. **545**, 487–498 (2021)
33. Frej, E.A., Roselli, L.R.P., Araújo de Almeida, J., de Almeida, A.T.: A multicriteria decision model for supplier selection in a food industry based on FITradeoff method. Math. Probl. Eng. **2017**, 1–9 (2017)

34. Santos, I.M., Roselli, L.R.P., da Silva, A.L.G., Alencar, L.H.: A supplier selection model for a wholesaler and retailer company based on FITradeoff multicriteria method. Math. Probl. Eng. **2020**, 8796282 (2020)

35. Dell'Ovo, M., Oppio, A., Capolongo, S.: Decision Support System for the Location of Healthcare Facilities Sit Health Evaluation Tool. Springer, Cham (2020). https://doi.org/10.1007/978-3-030-50173-0

36. e Silva, L.C., Daher, S.D.F.D., Santiago, K.T.M., Costa, A.P.C.S.: Selection of an integrated security area for locating a state military police station based on MCDM/A method. In: IEEE International Conference on Systems, Man and Cybernetics (SMC), Bari, Italy, pp. 1530–1534, October 2019

37. Camilo, D.G.G., de Souza, R.P., Frazão, T.D.C., da Costa Junior, J.F.: Multi-criteria analysis in the health area: selection of the most appropriate triage system for the emergency care units in natal. BMC Med. Inform. Decis. Mak. **20**(1), 1–16 (2020)

38. Shukla, S.: A fitradeoff approach for assessment and understanding of patient adherence behavior. In: Value in Health, vol. 20, no. 5, pp. A322. Elsevier Science Inc., New York, May 2017

39. de Morais Correia, L.M.A., da Silva, J.M.N., dos Santos Leite, W.K., Lucas, R.E.C., Colaço, G.A.: A multicriteria decision model to rank workstations in a footwear industry based on a FITradeoff-ranking method for ergonomics interventions. Oper. Res., 1–37 (2021)

40. Pergher, I., Frej, E.A., Roselli, L.R.P., de Almeida, A.T.: Integrating simulation and FITradeoff method for scheduling rules selection in job-shop production systems. Int. J. Prod. Econ. **227**, 107669 (2020)

41. Silva, M.M., de Gusmão, A.P.H., de Andrade, C.T.A., Silva, W.: The integration of VFT and FITradeoff multicriteria method for the selection of WCM projects. In: 2019 IEEE International Conference on Systems, Man and Cybernetics (SMC), 6–9 October, Bari, Italy, pp. 1513–1517 (2019)

42. Carrillo, P.A.A., Roselli, L.R.P., Frej, E.A., de Almeida, A.T.: Selecting an agricultural technology package based on the flexible and interactive tradeoff method. Ann. Oper. Res., 1–16 (2018)

43. Lima, E.S., Viegas, R.A., Costa, A.P.C.S.: A multicriteria method based approach to the BPMM selection problem. In: 2017 IEEE International Conference on Systems, Man, and Cybernetics (SMC), Banff, Canada, pp. 3334–3339, October 2017

44. de Gusmao, A.P.H., Pereira Medeiros, C.: A model for selecting a strategic information system using the FITradeoff. Math. Probl. Eng. **2016**(2), 1–7 (2016)

45. Fossile, D.K., Frej, E.A., da Costa, S.E.G., de Lima, E.P., de Almeida, A.T.: Selecting the most viable renewable energy source for Brazilian ports using the FITradeoff method. J. Clean. Prod. **260**, 121107 (2020)

46. Kang, T.H.A., Júnior, A.M.D.C.S., de Almeida, A.T.: Evaluating electric power generation technologies: a multicriteria analysis based on the FITradeoff method. Energy **165**, 10–20 (2018)

47. de Macedo, P.P., de Miranda Mota, C.M., Sola, A.V.H.: Meeting the Brazilian energy efficiency law: a flexible and interactive multicriteria proposal to replace non-efficient motors. Sustain. Cities Soc. **41**, 822–832 (2018)

48. Monte, M.B.S., Morais, D.C.: A decision model for identifying and solving problems in an urban water supply system. Water Resour. Manage **33**(14), 4835–4848 (2019)

49. da Silva, A.L.C.D.L., Costa, A.P.C.S., de Almeida, A.T.: Exploring cognitive aspects of FITradeoff method using neuroscience tools. Ann. Oper. Res., 1–23 (2021)

50. Silva, A.L.C.L; Costa, A.P.C.S.: FITradeoff decision support system: an exploratory study with neuroscience tools. In: NeuroIS Retreat 2019, Viena. NeuroIS Retreat (2019)

51. Roselli, L.R.P., Pereira, L., da Silva, A., de Almeida, A.T., Morais, D.C., Costa, A.P.C.S.: Neuroscience experiment applied to investigate decision-maker behavior in the tradeoff elicitation procedure. Ann. Oper. Res. **289**(1), 67–84 (2019). https://doi.org/10.1007/s10479-019-03394-w
52. Roselli, L.R.P., de Almeida, A.T.: Use of the Alpha-Theta Diagram as a decision neuroscience tool for analyzing holistic evaluation in decision making. Ann. Oper. Res. (2022)
53. Roselli, L.R.P., de Almeida, A.T.: The use of the success-based decision rule to support the holistic evaluation process in FITradeoff. Int. Trans. Oper. Res. (2021)
54. Pessoa, M.E.B.T., Roselli, L.R.P., de Almeida, A.T.: A neuroscience experiment to investigate the selection decision process versus the elimination decision process in the FITradeoff method. In: EWG-DSS 7th International Conference on Decision Support System Technology. Loughborough, United Kingdom (2021)
55. Reis Peixoto Roselli, L., de Almeida, A.: Analysis of graphical visualizations for multicriteria decision making in FITradeoff method using a decision neuroscience experiment. In: Moreno-Jiménez, J. M., Linden, I., Dargam, F., Jayawickrama, U. (eds.) ICDSST 2020. LNBIP, vol. 384, pp. 30–42. Springer, Cham (2020). https://doi.org/10.1007/978-3-030-462 24-6_3
56. Roselli, L., de Almeida, A.: Improvements in the FITradeoff decision support system for ranking order problematic based in a behavioral study with NeuroIS tools. In: Davis, F. D., Riedl, R., vom Brocke, J., Léger, P.-M., Randolph, A. B., Fischer, T. (eds.) NeuroIS 2020. LNISO, vol. 43, pp. 121–132. Springer, Cham (2020). https://doi.org/10.1007/978-3-030-60073-0_14
57. Roselli, L.R.P., de Almeida, A.T., Frej, E.A.: Decision neuroscience for improving data visualization of decision support in the FITradeoff method. Oper. Res. Int. J. **19**(4), 933–953 (2019). https://doi.org/10.1007/s12351-018-00445-1
58. Roselli, L., Frej, E., de Almeida, A.: Neuroscience experiment for graphical visualization in the FITradeoff decision support system. In: Chen, Y., Kersten, G., Vetschera, R., Xu, H. (eds.) GDN 2018. LNBIP, vol. 315, pp. 56–69. Springer, Cham (2018). https://doi.org/10.1007/978-3-319-92874-6_5
59. Roselli, L.R.P., de Almeida, A.T.: Behavioral study for holistic evaluation in FITradeoff method: hit rate for selecting versus eliminating alternatives. In: 21th International Conference on Group Decision and Negotiation in 2021, Toronto, Canada, GDN 2021, Proceedings (2021)
60. Rosch, J.L., Vogel-Walcutt, J.J.: A review of eye-tracking applications as tools for training. Cogn. Technol. Work **15**, 313–327 (2013)
61. Klimesch, W.: EEG alpha and theta oscillations reflect cognitive and memory performance: a review and analysis. Brain Res. Rev. **29**(2–3), 169–195 (1999)
62. Barla, S.B.: A case study of supplier selection for lean supply by using a mathematical model. Logist. Inf. Manag. **16**, 451–459 (2003)

Conflict Resolution

Meta Level Equilibrium Selection for Two Illustrative Noncooperative Games

Takahiro Suzuki[1]([⊠]) [ID] and Alexandre B. Leoneti[2] [ID]

[1] Department of Civil Engineering, The University of Tokyo,
7-3-1, Hongo, Bunkyo-ku, Tokyo, Japan
suzuki-tkenmgt@g.ecc.u-tokyo.ac.jp
[2] School of Economics, Business Administration and Accounting, University of São Paulo,
Ribeirão Preto, Brazil
ableoneti@usp.br

Abstract. Various equilibrium concepts such as Nash, general meta-rationality (GMR), sequential stability (SEQ), and so on, help anticipate different results of a conflict. In the Prisoner Dilemma (PD) game, for example, some players can choose not to cooperate (just as the Nash equilibrium anticipates) or they can choose to cooperate; perhaps this is because they fear others' counteraction (just as GMR anticipates). In any case, the actual players' choice is to select one of mode of rationality, such as Nash, GMR, or even other types of rationalities, on their own. The objective of this paper is to propose a new model that explains how such rationality is selected endogenously within the game (when such selection itself makes another game because the result depends also on other players' choice). Applying the convergence theory of procedural choice into the context of such equilibrium selection, we first model the meta-level selection of equilibria (which equilibrium players stand on to choose which equilibrium players stand on to choose their actions). Then, we apply the model to two typical games: PD game (cooperation matters) and Hi-Lo game (coordination matters). In both cases, we give a new explanation of why cooperation/coordination is possible within the conflicts.

Keywords: Rationality · Equilibrium selection problem · Graph model · Myopic versus nonmyopic

1 Introduction

The equilibrium selection problem is an important issue within game theory [1]. The problem is that a nonsingular set of outcomes is designated by the application of an equilibrium concept (e.g., the Nash equilibrium). Consequently, there is the necessity to select one equilibrium from among them—in some form. That selection usually requires an arbitration that is provided by exogenous information from the original game. In this sense, there are different approaches in the literature for selecting an equilibrium as the social outcome [2]. Given that its nature is based on an arbitration, there is no general method for this selection [3].

© The Author(s), under exclusive license to Springer Nature Switzerland AG 2022
D. C. Morais and L. Fang (Eds.): GDN 2022, LNBIP 454, pp. 61–73, 2022.
https://doi.org/10.1007/978-3-031-07996-2_5

This selection can be a special case of a more general problem of choosing a "principle" of behavior; either an equilibrium concept, policy, rationality, ground, or criterion; based on which players select their actions. We label this problem as a *general equilibrium selection problem*. Two typical examples of this are in order. The first one is the equilibrium selection problem itself, which was introduced above. In the two-player Stag Hunt game—for example—the criteria of risk-dominance (less risk is better) and payoff-dominance (more payoff is better) are two typical rationalities, with which players might choose their actions [4]. In other words, players are supposed to face the choice from these two (or even other) criteria when choosing the actual strategies of hunting a stag or hare. The second example is the choice from different equilibrium concepts. Consider the two-player Prisoner Dilemma (PD) game. As is well known, Nash stability (considering only one's unilateral improvements) and general meta-rationality (GMR) (considering the others' counteractions also) are two famous criteria; however, they can sometimes guide different actions (e.g., deviation from both players' cooperation is Nash stable but not GMR stable). This fact implies that players of the PD game should face the choice of level of rationality; i.e., to be myopic (without considering others' countermovements, just as Nash stability assumes) or not to be myopic (considering others' countermovements, just as GMR stability assumes).

Within social choice theory, a similar selection problem has been studied as the choice of voting rule in decision-making contexts. Among them, Suzuki and Horita [5] study how procedural choice can be made endogenously by the convergence of meta-level reasoning (which we call *convergence theory*). They assert that—from the original social choice problem—meta-level problems can be determined inductively, which often results in an eventual convergence of justifiable solutions. The basic idea is as follows: suppose that a society faces the choice problem X (the original set of alternatives), with F being the set of all feasible social choice correspondences (SCC). A preference profile L^0 over X is said to converge to $x \in X$ under the menu F if, when one sees a sufficiently high level "how to choose how to choose ... how to choose," every SCC in the menu ultimately results in the same alternative x. Technically speaking, a preference profile $L^0 \in \mathcal{L}(X)^n$ is said to *weakly converge* to $a \in X$ if there exists a consequentially induced (CI) sequence $L^0, L^1, L^2 \cdots L^{k-1}$; with which L^0 converges to a and L^0 is said to *strongly converge* to $a \in X$ if it weakly converges to a and never weakly converges to another alternative.

Within game theory, different studies have already been investigating the use of an equilibrium as the social outcome by means of the creation of meta-structures. For instance, Shubik [6] presented different approaches for this task, including the use of meta-games proposed by Nigel Howard [7]. Jehiel and Walliser [8] have demonstrated that a simple structure of a meta-game for selecting Nash equilibria within duopoly problems is possible by providing that the utility functions of the agents satisfy linear (or almost linear) constraints. Pozo et al. [9] used the approach of meta-games for providing to electricity companies the idea of having a range of offer strategies where several pure and mixed meta-game Nash equilibria are possible. Nevertheless, even with those attempts to solve the problem from its endogenous data, the creation of such meta-structures could also be considered an arbitrary procedure; this would lead us to the same arbitrariness of the exogenous approach.

On the other hand, unlike the arbitrariness of the procedures used for analyzing meta-structures in those studies, convergence theory does not stand on the view that certain equilibrium should be admitted *ex ante* at the meta-level. Rather, it simply focuses on what happens when the society regresses infinitely; i.e., considering how to choose how to choose... (and so on). This research proposes the application of the convergence theory for modeling and solving the *general equilibrium selection problem*.

To summarize, the present paper studies the *general equilibrium selection problem* based on endogenous data of the problem within a meta game in which the players are not deciding on the original strategies; instead, they are deciding on the way they should play the game. For example, the first level game could be the one in which either Nash or GMR is chosen by a player. The second level game, therefore, would be the one in which either Nash or GMR is chosen by a player for the first level game and so on. Higher level games are defined in the same way. Our study aims at modeling this meta-choice process and explain why a certain pathway, say every player follows GMR, is obtained as a result of endogenous meta-level structure (which we call convergence).

This construction—at first sight—may seem of low practical meaning because the troublesome infinite regress of how to choose how to choose (... and so on) just arises, but it is not. In the present paper, we show—by the perspective of the graph model for conflicts resolution (GMCR) [10]—that such construction of meta level games can converge in some cases. These include the PD game (a very known example of cooperation) and the Hi-Lo game (the very known example of coordination). This is possible since the meta-game proposes that the player will not decide about the strategies, but rather about the rationalities on which he or she will play the game.

Furthermore, it is particularly interesting that by solving the game from the means of convergence, the strategic choice of cooperation or coordination is also possible within the strategic interaction. According to Janssen [11], in a cooperation problem, as the PD, players have a dominant strategy, which is not to cooperate, but, eventually, players deviate from their dominant strategy and do cooperate, which could be explained by the axiom of individual rationality. However, still according to [11], in a coordination problem, the individual rationality is not sufficient to predict players' behavior, which would require an approach that supplements the traditional axioms of individual rationality. Therefore, by providing the interpretation of the strategic interaction in the perspective of a meta-structure, our approach intends to overcome the main limitation of the game theoretical models: their incapacity to clearly determine the results based simply on a material point of view [12].

The aim of this research is the proposition of a more general mechanism for the equilibrium selection, in the view of a *general equilibrium selection problem*. The proposed idea will be discussed in the light of the main concepts of equilibria for non-cooperative games. These include: (i) Nash Equilibrium [13]; (ii) General Metarationality (GMR) [7]; (iii) Symmetric Metarationality (SMR) [7]; (iv) Sequential Stability (SEQ) [14]; (v) Stability of limited movements (Lh) [15, 16]; and (vi) Non-myopic stability (N-M), a particular case of Limited Movement Stability (Lh) [17]. The choice of non-cooperative games' solution concepts of pure strategies is because such structure is similar to that considered in social choice [18–20].

The next section presents the methodology, followed by sections covering the results and discussions of this study. Finally, the last section presents the conclusion of this research.

2 Model

2.1 Definition of a Graph Model

A directed graph $G = (S, T)$ consists of the set S of nodes and the set T of edges. A directed graph G is called simple when there is no loop or multiple edge. For two simple directed graphs $G_1 = (S_1, T_1)$ and $G_2 = (S_1, T_2)$, if they have the same set of nodes S_1, their *sum graph*—denoted by $G_1 + G_2$—is defined as $G_1 + G_2 := (S_1, T_1 \cup T_2)$. The *sum graph* of more than two graphs is defined in the same way. For simple directed graphs $G_1 = (S_1, T_1), G_2 = (S_1, T_2), \cdots, G_n = (S_1, T_n)$ (with the same set of nodes S_1), their *sum graph* is defined as $\sum_{i \in N} G_i := \left(S_1, \bigcup_{i=1}^{n} T_i\right)$.

Next, graph model is defined. Among the vast literature on GMCR, our definition basically follows [10]. A *graph model* $\mathcal{M} = < N, S, (G_i)_{i \in N}, (\succsim_i)_{i \in N} >$ consists of

- a finite set of players $N = \{1, 2, \cdots, n\}$
- a finite set of states S
- a simple directed graph $G_i = (S, T_i)$ for each $i \in N$
- weak preference $\succsim_i$ over S for each $i \in N$.

We call $G_i = (S, T_i)$ as *player i's graph at the graph model* $\mathcal{M}$. It represents the unilateral move that player i can conduct alone. For two states $s, s' \in S$, an edge $(s, s') \in T_i$ means that player i can move from state s to state s' by oneself. For $s \in S$ and $i \in N$, we denote as $R_i(s) := \{t \in S \mid (s, t) \in T_i\}$ (i's reachable list from s) and $R_i^+(s) := \{t \in R_i(s) \mid t \succ_i s\}$ (i's unilateral improvement list from s).

2.2 Equilibrium Selection Problem

In the literature, several equilibrium concepts have been proposed including Nash, GMR, SEQ, L_h, and so on. In this article, we regard each of them as a principle that guides players to take certain movements[1]. For an equilibrium concept X = Nash, GMR, SEQ, $L_h, \cdots$, a state s is called X-stable for player i if the player has no incentive to move unilaterally from s according to the principle of X. A state s is called an X-equilibrium if it is X-stable for all players.

One problem arises when different equilibria concepts designate different outcomes. Suppose that a state s is Nash-stable for a player but s is not SEQ-stable for the same player. In such a case, how can the player select one from Nash and SEQ? This choice, which we call a *general equilibrium selection problem*, can also be a game because i's payoff should be dependent on others' choices.

---

[1] It is worth noting that we focus on the functional aspect of equilibria. In the perspective of GMCR theory, an equilibrium is a *state* that is stable for all players. As we noted in the introduction, we regard equilibria as principles (or rationalities), with which players choose their actions.

In general, starting from the original graph model $\mathcal{M}$, the players' first choice should be which of Nash, GMR, SEQ, L_h, $\cdots$ they should follow to select their movements in $\mathcal{M}$. This makes another graph model: $\mathcal{M}^1$. However, since players—again—require certain principles to solve this graph model, they soon face the second level graph model: $\mathcal{M}^2$ (i.e., which of Nash, GMR, SEQ, L_h, $\cdots$ should they follow to select their movements in $\mathcal{M}^1$). Accordingly, this inference can go on ad infinitum ($\mathcal{M}^3, \mathcal{M}^4, \cdots, \mathcal{M}^k, \cdots$).

The contribution of this paper is to show that this inference is not meaningless.

2.3 Meta-structures Derived from Equilibria

Formally, let $\mathcal{E} \subseteq \{$Nash, GMR, SEQ, $L_h\}$ be the set of all $e_i \in \mathcal{E}$ equilibria that are considered. Elements of $\mathcal{E}$ will be typically called *principles* (rationalities, policies, criteria, etc.). We say that a state t is a Nash-improvement of state s for player i if $t \in R_i^+(s)$. State t is said to be a GMR-improvement of state s for player i if $t \in R_i^+(s)$ and there is no $u \in R_j(t)$ such that $u \succsim_i s$. State t is said to be a SEQ-improvement of state s for player i if $t \in R_i^+(s)$ and there is no $u \in R_j^+(t)$ such that $u \succsim_i s$. Finally, the concept of limited move improvement/stability is defined through an extensive form game (see, e.g., [10] for a formal definition). Player i's game with initial state s is an extensive form game with complete information that begins from s and continues until either when (i) either player chooses to stay, or (ii) h moves are consecutively chosen. State t is said to be a L_h-improvement of state s for player i if the movement (s, t) is obtained through the backward induction of that game.

Whether a certain unilateral movement (s, t) is "rational" or not for a player depends on which principle in $\mathcal{E}$ the player selects. A player with Nash should take any unilateral improvements, while a player with GMR would not take those unilateral improvements if there is a fear of counteraction by their opponents.

Definition 1 (player i's e_i-graph at $\mathcal{M}$)
Let $\mathcal{M} = < N, S, (G_i)_{i \in N}, (\succsim_i)_{i \in N} >$ be a graph model, $i \in N$, and $e_i \in \mathcal{E}$. We define i's e_i-graph at $\mathcal{M}$, denoted by graph$[i, e_i, \mathcal{M}] := (S, T_i)$, as a directed graph such that.

1) the set of nodes S is the set of states in $\mathcal{M}$.
2) the set of edges T_i is defined as follows: for each $s, s' \in S$, $[(s, s') \in T_i$ if and only if s' is an e_i-improvement of s for player $i]$.

Definition 2 (e-graph at $\mathcal{M}$)
For $e = (e_1, e_2, \cdots, e_n) \in \mathcal{E}^n$, we define *e-graph at $\mathcal{M}$*, denoted by graph$[e, \mathcal{M}]$, as

$$\text{graph}[e, \mathcal{M}] := \sum_{i \in N} \text{graph}[i, e_i, \mathcal{M}].$$

2.4 Domination Between Graphs

For any simple directed graph $G = (S, T)$ and its node $s \in S$, we define the *destination of s in G*, denoted by $\varphi_G(s)$, as the set of nodes to which s ultimately reaches; i.e.,

$$\varphi_G(s) := \{t \in S (\exists \text{path from } s \text{ to } t \text{ in } G) \,\&\, (\neg\exists \text{edge from } t \text{ in } G)\}.$$

Suppose that two graphs $G = (S, T)$ and $G' = (S, T')$ have the same set of nodes and each node has nonempty destination. Let $\succsim_i$ be a weak preference on S. We say that G *dominates* G' with respect to $\succsim_i$ if for all $s \in S$, any element of $\varphi_G(s)$ is at least as good as any element of $\varphi_{G'}(s)$; i.e.,

$$t \succsim_i u \text{ for all } t \in \varphi_G(s) \text{ and } u \in \varphi_{G'}(s).$$

In addition, we say that G *strongly-dominates* G' with respect to $\succsim_i$ if G dominates G' with respect to $\succsim_i$ and there exists $s \in S$, $t \in \varphi_G(s)$ and $u \in \varphi_{G'}(s)$, such that $t \succ_i u$.

2.5 Meta Level Choice

Suppose that the society is equipped with the *original graph model* $\mathcal{M}^0 = \, < N, S^0, (G_i^0)_{i \in N}, (\succsim_i^0)_{i \in N} >$, which is alternatively called as the *level-0 graph model*. We are now ready to formulate the scenario stated in Subsect. 0; i.e., to define the meta level graph models $\mathcal{M}^1, \mathcal{M}^2, \cdots$, which are generated by $\mathcal{M}^0$.

The construction is made inductively. Suppose that we have $\mathcal{M}^0, \mathcal{M}^1, \cdots, \mathcal{M}^{k-1}$ for some positive integer $k \in \mathbb{N}$. *Level-k graph model* $\mathcal{M}^k$ is defined as $< N, \mathcal{E}^n, G_i^k = (\mathcal{E}^n, T_i^k), (\succsim_i^k)_{i \in N} >$, where.

(i) In player i's graph $G_i^k = (\mathcal{E}^n, T_i^k)$, an edge from $(e_1, e_2, \cdots, e_n) \in \mathcal{E}^n$ to $(e_1', e_2', \cdots, e_n') \in \mathcal{E}^n$ exists in T_i^k if and only if $e_i \neq e_i'$ and $e_j = e_j'$ for all $j \neq i$.

(ii) Player i's weak preference $\succsim_i^k$ satisfies the followings.

(a) If graph$[e, \mathcal{M}^{k-1}]$ dominates graph$[e', \mathcal{M}^{k-1}]$ with respect to $\succsim_i^{k-1}$, then $e \succsim_i^k e'$.

(b) If graph$[e, \mathcal{M}^{k-1}]$ strongly-dominates graph$[e', \mathcal{M}^{k-1}]$ with respect to $\succsim_i^{k-1}$, then $e \succ_i^k e'$.

Definition 3 (convergence)

We say that level-0 graph model $\mathcal{M}^0 = < N, S, (G_i)_{i \in N}, (\succsim_i)_{i \in N} >$ *converges* under $\mathcal{E}$ if there exists $k \in \mathbb{N}$, such that graph$[e, \mathcal{M}^k]$ are the same for all $e \in \mathcal{E}^n$.

Several comments are in order. First, given $\mathcal{M}^0, \mathcal{M}^1, \cdots, \mathcal{M}^{k-1}$, level-$k$ graph model $\mathcal{M}^k$ is not unique in general; this is because the above definition says nothing if there is no dominance relation between e and e' (i.e., if none of graph$[e, \mathcal{M}^{k-1}]$

and graph$[e', \mathcal{M}']$ dominate the other). In this sense, our definition stipulates the class of possible $\mathcal{M}^k$ in the face of lower-level graph models $\mathcal{M}^0, \mathcal{M}^1, \cdots, \mathcal{M}^{k-1}$ (Nevertheless, this indeterminacy of meta-level preferences does not matter in the two games discussed below). Second, our construction of $\mathcal{M}^0, \mathcal{M}^1, \mathcal{M}^2, \cdots$ is "endogenous." As is shown in the definition, all one needs to know to define (the possible class of) $\mathcal{M}^k$ is the lower-level graph models $\mathcal{M}^0, \mathcal{M}^1, \cdots, \mathcal{M}^{k-1}$—as well as the domain of principles $\mathcal{E}$. Our model does not make further assumptions on players' choice of principles (which combination $e \in \mathcal{E}^n$ players actually follow). This inductive construction of meta structures is similar to Suzuki and Horita [5], which constructs meta level decision-making problems inductively based on the structure of lower levels.

3   Diagram Expression of Graph Models

In the meta-level arguments on graph models, we often need to mention multiple graph models at the same time (the original graph model, level-1 graph model, level-2 graph model, and so on). To simplify this, we introduce a visual way to express a graph model by way of a single diagram. The diagram is obtained by applying the graph representation by Shubik [10, Fig. 1] into our contexts of graph models.

Fix $N = \{1, 2\}$. *Diagram expression* of a graph model $\mathcal{M} = \, < N, S, (G_i)_i, \left(\succsim_i\right)_i >$ is to represent $\mathcal{M}$ by a single diagram[2]. Each node is an outcome in S, an edge from s to t with label i means that (s, t) is one of player i's unilateral movements, and a state s is located in the east (resp. in the north) of state t if and only if s is preferred to t by player 1 (resp. player 2).

Let us demonstrate this method by an example. Suppose that $S = \{a, b, c\}$ (three outcomes), $G_1 = (S, T_1)$ with $T_1 = \{(a, b)\}$ (i.e., player 1 can move only from a to b), $G_2^0 = (S, T_2)$ with $T_2 = \{(b, c)\}$ (i.e., player 2 can move only from b to c), $b \succ_1^0 a \succ_1^0 c$ (player 1's preference), and $b \sim_2^0 a \succ_2^0 c$ (player 2's preference). Its diagram expression is shown in Fig. 1.

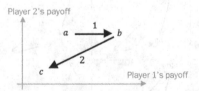

Fig. 1. Diagram expression of $\mathcal{M}$.

Note that the diagram fully contains the information in the formal description of $\mathcal{M}$ (if the set of players N is explicit). We can read $S = \{a, b, c\}$ by the nodes in the diagram, $G_i = (S, T_i)$ by the edges with label i, and the preference $\succsim_i$ by the positions of the nodes. Therefore, we can represent a graph model by such a diagram expression instead of determining $\mathcal{M} = \, < N, S, (G_i)_i, \left(\succsim_i\right)_i >$ with formulae.

---

[2] Technically speaking, diagram expression is more than what is called a (directed) graph in graph theory; this is because we also care about the positions of nodes, in addition to the set of nodes and set of edges. Therefore, we call it diagram expression rather than graph expression.

4 Numerical Examples

We discuss PD game under $\mathcal{E} = \{\text{Nash}, \text{L}_2\}$. Beginning from the original game $\mathcal{M}^0$ (Fig. 2), we demonstrate that convergence is found at the third level.

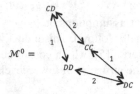

$$\mathcal{M}^0 =$$

Fig. 2. Diagram expression of PD (we denote by XY the outcome such that player 1 selects strategy $X = C, D$ and player 2 selects strategy $Y = C, D$).

Meta level graph models $\mathcal{M}^1, \mathcal{M}^2, \cdots$ are defined in the following way.

- On $\mathcal{M}^1$.

Among 1's unilateral movements, (CC, DC) and (CD, DD) are Nash improvements and only (CD, DD) is the unique L_2-improvement. Therefore, graph$[e, \mathcal{M}^0]$'s is shown in Table 1, where the cell of row X and column Y represents graph$[(X, Y), \mathcal{M}^0]$.

Table 1. Graph $[e, \mathcal{M}^0]$'s for PD game.

	Nash	L_2
Nash		
L_2		

One can see that under each $e = (\text{Nash}, \text{Nash}), (\text{Nash}, L_2), (L_2, \text{Nash})$, all nodes in $\mathcal{E}^n$ have destination $\{DD\}$ (i.e., under these three realizations of players' attitudes, beginning from any outcome, the game must result in DD). On the other hand—under $e = (L_2, L_2)$—the destination of CC is $\{CC\}$, while the destinations of the other nodes remain $\{DD\}$. Therefore, it follows that:

(i)   Three graphs graph[(Nash, Nash), $\mathcal{M}^0$], graph[(Nash, L$_2$), $\mathcal{M}^0$], and graph[(L$_2$, Nash), $\mathcal{M}^0$] dominate each other with respect to both $\succsim_1^0$ and $\succsim_2^0$. Therefore, the three outcomes (Nash, Nash), (Nash, L$_2$), (L$_2$, Nash) are judged indifferent in $\mathcal{M}^1$ by both $\succsim_1^1$ and $\succsim_2^1$.

(ii). The three graphs are strongly dominated by graph[(L$_2$, L$_2$)], with respect to both $\succsim_1^0$ and $\succsim_2^0$. So, the outcome (L$_2$, L$_2$) is preferred to (Nash, Nash), (Nash, L$_2$), and (L$_2$, Nash) in $\mathcal{M}^1$ by both $\succsim_1^1$ and $\succsim_2^1$. Now, we obtain the diagram expression of $\mathcal{M}^1$ (Fig. 3).

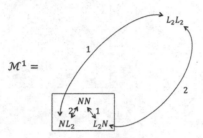

Fig. 3. $\mathcal{M}^1$ for PD game (three outcomes in the rectangle are indifferent for both players).

- On $\mathcal{M}^2, \mathcal{M}^3, \cdots$.

Just as we constructed $\mathcal{M}^1$ from $\mathcal{M}^0$, we can construct $\mathcal{M}^2, \mathcal{M}^3, \cdots$ inductively. Here are the results (Fig. 4):

Table 2. Graph $\left[e, \mathcal{M}^1\right]$'s for PD game.

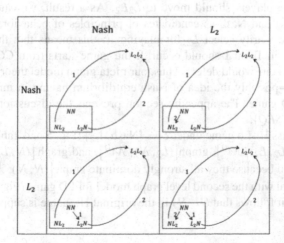

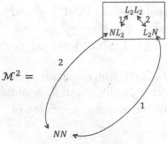

$$\mathcal{M}^2 =$$

Fig. 4. $\mathcal{M}^2$ for PD game

Table 3. Graph $\left[e, \mathcal{M}^2\right]$'s for PD game.

	Nash	L_2
Nash	(graph)	(graph)
L_2	(graph)	(graph)

One can see that for each $e \in \mathcal{E}^2$, graph$[e, \mathcal{M}^2]$ is the same. This means that the original PD game converges.

Let us detail the interpretation of this result. According to Table 3, no matter which of (Nash, Nash), (Nash, L_2), (L_2, Nash), or (L_2, L_2) players select as their principle of behavior for second level graph model $\mathcal{M}^2$, the resulting graph$[e, \mathcal{M}^2]$ is the same. This result has two messages. One is that further regression of $\mathcal{M}^3$, $\mathcal{M}^4$, $\cdots$ is supposed to have no effective meaning; this is because no matter which element $e \in \mathcal{E}^n$ turns out to be appropriate for level-3 or higher meta-level selection, the result would not change at all. The other message is that either one of L_2L_2, NL_2, or L_2N as states in $\mathcal{M}^2$ should realize; i.e., at least one of the players should select L_2 rather than Nash for playing $\mathcal{M}^1$.

According to Table 2, it follows that only L_2L_2 in $\mathcal{M}^1$ is stable; starting from any other outcome, the players should move to L_2L_2. As a result, we can say that such players—having L_2 and Nash as candidates of principles of behavior in playing PD game—should necessarily select L_2 for playing it; this means that the cell of row-L_2 and column-L_2 in Table 1 should occur. If the game starts from CC (both players cooperate), then no one would defect. Thus, our meta graph model theory explains how a rational player—possibly the idea of Nash equilibrium having in mind—can make cooperation in PD game. The appendix section presents the discussions of PD game under $\mathcal{E} = \{\text{Nash}, SEQ\}$.

Now, we discuss Hi-Lo game under $\mathcal{E} = \{\text{Nash}, L_2\}$ (Fig. 5 and Table 4).

Since graph$[(L_2, L_2), \mathcal{M}^0]$, graph$[(L_2, N), \mathcal{M}^0]$, and graph$[(N, L_2), \mathcal{M}^0]$ dominate each other and because they all strongly dominate graph $[(N, N), \mathcal{M}^0]$, it follows that $\mathcal{M}^1$ is identical with the second level graph model for PD game (Fig. 4). In a similar vein as PD game, it follows that (H, H) in the original HL game is supposed to realize.

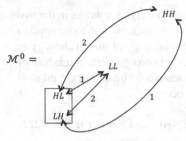

Fig. 5. $\mathcal{M}^0$ for HL game

Table 4. Graph $\left[e, \mathcal{M}^0\right]$'s for HL game.

5 Final Remarks

In this article, we proposed a way to solve the *general equilibrium selection problem* by means of a meta-level structure; in which each player is supposed to select an equilibrium as their principle of behavior and that such a selection itself makes another game. The proposition of the meta-level structure is similar to the one proposed by Suzuki and Horita [5] for the scenario of social choice theory. Although the idea of solving the equilibrium selection problem by means of meta-structures is not new; for e.g., [6–9]; such approaches also consider arbitrary procedures for setting such meta-structures. On the other hand, our approach minimizes such arbitrariness using the convergence theory, which establishes—for some situations as shown in the present paper (e.g., $\mathcal{E} = \{\text{Nash}, L_2\}$)—a stable solution to the equilibrium selection problem. It is particularly interesting to note that this new approach can be used for explaining how cooperation can be realized in PD game. Furthermore, classic studies on graph models for conflict resolutions, including metagame theory, deal with different equilibria independently; furthermore, they assume that the modeler or analyst selects appropriate equilibrium for the context. On the other hand, our model directly analyzes how each player endogenously selects an appropriate equilibrium.

Finally, it is worth noting that whether a certain game (say, PD game) converges or not surely depends on the domain of equilibria $\mathcal{E}$ considered. Suppose $\mathcal{E} = \{\text{Nash}\}$, as a trivial case. Then, the game converges at the original level and the upper left scenario of Table 1 is obtained (any state must proceed to "DD"). Sect. 0 shows that adding L_2; i.e., considering $\mathcal{E} = \{\text{Nash}, L_2\}$; allows the society the opportunity to find the scenario of stable cooperation. On the other hand, if SEQ is considered instead of L_2, the cooperation scenario might not be found (See Appendix). A possible explanation for this difference is that players are too myopic to find improvements that take two steps under $\mathcal{E}' = \{\text{Nash}, \text{SEQ}\}$. This partly explains why cooperation is feasible in some cases but not in other cases. The present paper shows how the idea of convergence works in the two illustrative games (PD and HL), but our model in Sect. 0 is general enough to be applied to cases with more players, strategies, or criteria (with the analogy of mixed strategy in the standard game theory, one can even imagine a player following a certain mixture of Nash and GMR, for example). Investigation of these cases would be an important next step to make clear when and how convergence actually occurs.

Acknowledgment. This work was supported by JSPS KAKENHI Grant Number JP21K14222.

Appendix (the Case of $\mathcal{E}' = \{$Nash, SEQ$\}$ in PD Game)

The tables below show how the PD game converges under $\mathcal{E}'$. By considering graph$\left[e, \mathcal{M}^1\right]$'s, one can see that every realization of $e \in \mathcal{E}^n$ gives the same graph$\left[e, \mathcal{M}^1\right]$ (convergence). One major difference between Table 2 and Table 5 is, however, that NN is a fixed point in the latter (the destination of NN is $\{NN\}$ itself). As a result, once both players follow "N", the graph model will result in graph$\left[(\text{Nash, Nash}), \mathcal{M}^0\right]$ (upper left case in the left table of Table 5). Therefore, any state must result in "DD" and players—consequently—never find a way of cooperation. The reason for this failure is that no element in $\mathcal{E}' = \{$Nash, SEQ$\}$ can find a deviation from "NN" in graph$\left[e, \mathcal{M}^1\right]$ (right of Table 5).

Table 5. Graph$\left[e, \mathcal{M}^0\right]$'s (left) and graph$\left[e, \mathcal{M}^1\right]$'s (right) for PD game with $\mathcal{E}'$

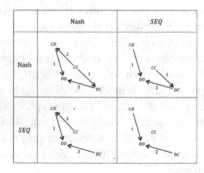
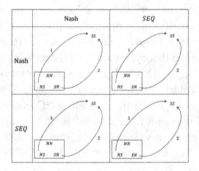

References

1. Harsanyi, J.C., Selten, R.A.: General Theory of Equilibrium Selection in Games, MIT Press Books (1988)
2. Leoneti, A.B., Prataviera, G.A.: Entropy-norm space for geometric selection of strict Nash equilibria in n-person games. Physica A **546**, 124407 (2020)

3. Mailath, G.J.: Do people play Nash equilibrium? Lessons from evolutionary game theory. J. Econ. Lit. **36**(3), 1347–1374 (1998)

4. Harsanyi, J.C.: Papers in Game Theory. Springer, Dordrecht (1982). https://doi.org/10.1007/978-94-017-2527-9

5. Suzuki, T., Horita, M.: Plurality, borda count, or anti-plurality: regress convergence phenomenon in the procedural choice. In: Bajwa, D., Koeszegi, S.T., Vetschera, R. (eds.) GDN 2016. LNBIP, vol. 274, pp. 43–56. Springer, Cham (2017). https://doi.org/10.1007/978-3-319-52624-9_4

6. Shubik, M.: Game theory, behavior, and the paradox of the prisoner's dilemma: three solutions. J. Conflict Resolut. **14**(2), 181–193 (1970)

7. Howard, N.: Paradoxes of Rationality: Theory of Metagames and Political Behavior. MIT Press, Cambridge (1971)

8. Jehiel, P., Walliser, B.: How to select a Dual Nash equilibrium. Games Econom. Behav. **10**(2), 333–354 (1995)

9. Pozo, D., Contreras, J., Caballero, Á., de Andrés, A.: Long-term Nash equilibria in electricity markets. Electr. Power Syst. Res. **81**(2), 329–339 (2011)

10. Kilgour, D.M., Hipel, K.W., Fang, L.: The graph model for conflicts. Automatica **23**(1), 41–55 (1987)

11. Janssen, M.C.: Coordination and cooperation. Behav. Brain Sci. **26**(2), 165–166 (2003)

12. Graziano, M.: Individual and social preferences: defending the agent's perspective rather than the theoretician's. Philos. Soc. Sci. **45**(2), 202–226 (2015)

13. Nash, J.: Non-cooperative games. Ann. Math. **54**(2), 286–295 (1951)

14. Fraser, N.M., Hipel, K.W.: Solving complex conflicts. IEEE Trans. Syst. Man Cybern. **9**(12), 805–816 (1979)

15. Zagare, F.C.: Limited-move equilibria in 2×2 games. Theor. Decis. **16**(1), 1–19 (1984)

16. Fang, L., Hipel, K.W., Kilgour, D.M.: Interactive Decision Making: The Graph Model for Conflict Resolution. Wiley, New York (1993)

17. Brams, S.J., Wittman, D.: Nonmyopic equilibria in 2×2 games. Confl. Manag. Peace Sci. **6**(1), 39–62 (1981)

18. Ziotti, V.C., Leoneti, A.B.: Improving commitment to agreements: the role of group decision-making methods in the perception of sense of justice and satisfaction as commitment predictors. Pesquisa Operacional **40** (2020)

19. Abreu, D., Sen, A.: Virtual implementation in Nash equilibrium. Econometrica J. Econ. Soc., 997–1021 (1991)

20. Maskin, E.: Nash equilibrium and welfare optimality. Rev. Econ. Stud. **66**(1), 23–38 (1999)

Conflict in Tiny Town: Aggregate Mining at the Alliston Aquifer

Simone Philpot[1]([✉]), Nayyer Mirnasl[1], and Keith W. Hipel[1,2]

[1] Department of Systems Design Engineering, University of Waterloo, Waterloo, ON, Canada
sphilpot@uwaterloo.ca
[2] Centre for International Governance Innovation, and Balsillie School of International Affairs,
University of Waterloo, Waterloo, ON, Canada

Abstract. Conflict over an application to expand aggregate mining in the Township of Tiny, Ontario, is investigated using the Graph Model for Conflict Resolution. This analysis sheds light on a consequential and growing class of multi-jurisdictional resources conflict in Ontario, aggregate mining. At the same time, because the Alliston aquifer contains the purest known naturally occurring water globally, this work describes an aggregate dispute in a unique setting. Analysis indicates that the evaluation process provides more opportunities to approve the application to expand mining at the site. However, some residents opposing the site are seeking novel ways to intervene in the process, such as appealing for federal intervention based on the unique status of the aquifer.

Keywords: Aggregate mining · Water resources · Graph Model for Conflict Resolution · Township of Tiny

1 Introduction

Residents in the Township of Tiny are no strangers to conflict. Situated atop the Alliston aquifer this region contains uniquely pristine groundwater. Many residents of Tiny and nearby communities have campaigned for decades against land uses they deemed to carry too high a risk to this unique resource [1]. An application to expand existing aggregate mining activities at French's Hill, where the aquifer is recharged, has mobilized community organizations once again [2–6]. We investigate conflict arising over this application, known as the Teedon Pit expansion, using the Graph Model for Conflict Resolution (GMCR) [7–11].

This work contributes to an understanding of the current situation in aggregate mining management in Ontario, and the conflicts that arise from it. Moreover, we investigate a unique setting with an unparalleled quality of water, over which preferences around competing uses, values, and priorities are currently being negotiated.

In Sect. 2, we describe aggregate mining and its relevance to conflict. Section 3 provides information about the study area. We then describe our analytical method, the Graph Model for Conflict Resolution (GMCR) [7–11], in Sect. 4. In Sect. 5 we present the Teedon expansion conflict model. Stability results are summarized in Sect. 6, followed by a brief discussion, and concluding remarks in Sect. 7.

© The Author(s), under exclusive license to Springer Nature Switzerland AG 2022
D. C. Morais and L. Fang (Eds.): GDN 2022, LNBIP 454, pp. 74–90, 2022.
https://doi.org/10.1007/978-3-031-07996-2_6

2   Aggregate Mining and Conflict

Aggregates is a term describing a variety of earth materials including gravel, sand, clay, limestone, marble, granite, and other similar goods [12]. Aggregate extraction is the act of mining those materials to put them to beneficial uses as critical inputs for infrastructure, roads, and many other products that are foundational to urban living [13].

Aggregate extraction has some defining features shaping the conflicts arising from it. First, as a point source resource required for urban development, harms associated with aggregate mining are experienced locally while the benefits of aggregate use are distributed across society, at national and provincial scales, and to the aggregate industry [13–15]. In Canada, this brings into conflict the needs of communities where aggregates are located with the priorities held by provincial governments and by businesses who own the contested lands [16]. Further complicating this area of decision-making, aggregate sites frequently overlap with other highly valued lands such as productive agricultural soils and high-quality groundwater [16]. This gives rise to difficult trade-offs between critical resources overlapping in place. Finally, while there are publicly owned aggregate mines operated on Crown Land, most sites in Ontario, Canada are on private lands [16]. As such, aggregate conflicts intersect with political, economic, and social norms over private property rights and the public interest [16]. Altogether, aggregate mining in Ontario is a special case of conflict bringing together multiple jurisdictions from public and private spheres in competition over how to prioritize and manage critical resources.

3   Study Area

3.1   Aggregate Governance in Ontario

In Canada, aggregate mining is largely governed by provincial governments [17]. In Ontario, the process involved in applying to excavate aggregates varies depending on a range of factors including the type of land (private ownership or Crown land), the depth of extraction, the amount of aggregate to be extracted, and other details specific to a given application [12]. Herein, our description of aggregate governance will focus solely on sites on private land in an Ontario context, as is the case in the Teedon expansion application.

The Ministry of Northern Development, Mines, Natural Resources and Forestry (MNDMNRF) has the power and responsibility to issue or deny licenses for aggregate mining [12]. However, many responsibilities related to evaluating and approving an aggregate application are distributed to regional and local municipalities. These local government entities implement aspects of the governing legislation, public participation, oversight, and ensure that the application aligns with regional and municipal planning documents. Moreover, local governments control bylaws and the zoning of lands. They are required to approve or reject applications to designate lands to a use compatible with aggregate activities if needed. That said, the Provincial government has the power to issue Minister's Zoning Orders (MZOs) that unilaterally bypass local decisions and public participation to permit land uses that would otherwise contravene existing plans [12, 18, 19].

3.2   Township of Tiny

Approximately 12,000 people live permanently in The Township of Tiny [20]. It is located on 335 square kilometers in south-central Ontario, in the most Northerly portion of Simcoe County. Its neighbors include the Township of Tay, Springwater, and seventy kilometers of the Georgian Bay coastline (Fig. 1) [21]. About 80% of the households in Tiny rely on well water [22]. Groundwater in the region is abundant and an important water source for the Wyte River and the Georgian Bay [23]. The region's high quality and plentiful groundwater is a source of community pride, with an annual water festival celebrating an artesian well located in the town center [6].

The Township already hosts the extractive industry, including, for example, gravel quarries in the area that have been operating for more than a decade and two aggregate wash water facilities [5, 24]. In 2009 as extraction at one site expanded residents began reporting changes to the quality of their well water. While some residents and experts have attributed these water quality problems to gravel washing conducted for aggregate operations, the company and the Ontario Ministry of the Environment argue that they are caused by poor well construction or maintenance [6, 22].

3.3   The Alliston Aquifer

The Teedon pit expansion would run below the Alliston aquifer, a potentially unique feature that may contain the world's cleanest natural water [3, 4, 23]. The water produced by the geology present at the Alliston aquifer was once thought to contain ancient water, but studies indicate it contains waters that rained down in recent decades, not centuries. This means that the quality of the water is maintained by the geologic conditions overlaying it, actively removing pollutants, and protecting the deposits. Because of its purity, the Alliston aquifer waters are already being used as a benchmark for water quality studies and researchers have proposed a study to determine how the site produces such pure water [1, 3, 4, 23]. Residents are concerned that expanded aggregate operations will threaten this unique feature, further burden the area, and raise the risk of well water contamination. Researchers have asked for five years to study the aquifer before licensing the Teedon pit extension [3, 6, 25, 26].

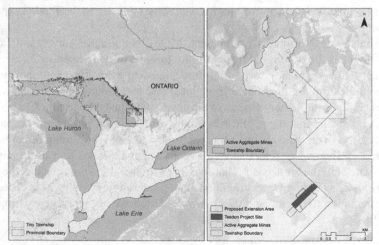

Fig. 1. Map of the study area

3.4 The Teedon Pit Expansion Application

Efforts to expand aggregate mining in the area are complicated, involving separate applications for interdependent aspects of the proposed project. For example, the company applying to expand their operations, CRH, is engaged in a separate but related application process to request a new water taking permit that would be needed to support the expansion [2]. While aspects of this water permit application are related to the Teedon Pit expansion, they are submitted independently. This conflict model will focus on the process involved in applying for the Teedon Pit Expansion under the Aggregate Resources Act.

CRH, acting through its subsidiary Dufferin Aggregates', is applying under the Aggregates Resources Act for a Category 3, Class A license for above water excavation to a maximum of 600,000 tonnes of aggregate per year. The actual amount of aggregate licensed to be excavated from the existing site is not changed by this expansion, as the tonnage to be removed is combined with the original license which is already approved for that amount [27, 28]. If approved, a license would expand their Teedon Pit extraction northward adding 15.3 hectares to their licensed area, 13.5 hectares of which would be proposed for extraction [27, 28]. The site under consideration is located on the traditional lands of the Anishinabe people of Beausoleil First Nation. It is currently designated as 'Greenlands' and 'Rural' in the Simcoe County Official Plan and as 'Mineral Aggregate Resources Two' and 'Environmental Protection Two' in the Township of Tiny Official Plan. The Township has the parcel of land zoned as 'Rural'. To receive a license for the Pit expansion, the proponent will need a Township of Tiny Official Plan Amendment and a Zoning By-law Amendment. They will then require a license from the MNDMNRF [27].

Originally initiated in 2012, the Teedon Pit expansion was subject to lengthy negotiations, gaining momentum only after the land was purchased by CRH in 2017 [2]. In 2019, the Township of Tiny opposed an application from CRH at that time, issuing a statement indicating their commitment to protecting the water in the region and allowing time for further research at the aquifer, as well as a range of technical concerns regarding the application [29]. However, they then reversed that decision in May of 2021 arguing that CRH had addressed their technical concerns [30]. The CRH license proposal was posted to the Environmental Registry of Ontario on October 10, 2019 [28]. At the time of this analysis, the comment period for the application had ended but no decision had been issued by the MNDMNRF.

4 Methodology

The Teedon pit expansion was investigated using the Graph Model for Conflict Resolution (GMCR) [7–11], a game theoretic approach distinguished, in part, by its use of ordinal preferences and its compatibility with a suite of solution concepts for describing human behavior under conflict. With this flexible method, analysts create robust models of conflict situations even when preference information is limited. Moreover, decision-makers with different levels of foresight, knowledge, and risk aversion can be considered.

GMCR has been used to investigate a wide range of conflicts including, but not limited to, water resources management [31–37], aquaculture [38, 39], Brownfields [40–46] and energy [47, 48].

GMCR provides a systematic approach to defining a given conflict. After identifying a conflict and moment in time for analysis the information required includes the decision-makers, a set of options each decision-maker can unilaterally control, a set of feasible states that could occur in the conflict, and a ranking of those feasible states from most to least preferred for each decision-maker [49]. Each state is then examined in terms of its stability [50]. For stability analysis, it is critical to understand the concepts of unilateral moves (UMs) and unilateral improvements (UIs). If a given decision-maker can, through their own unilateral action, move the conflict from one state to another state they are said to have a unilateral move. If that unilateral move results in the conflict reaching a state that is more preferred than the initial state, it is then defined by the more restricted concept of unilateral improvement. These concepts are then foundational to a range of stability definitions that take into account different features of decision-makers. For example, Nash stability [51, 52] defines a given state as stable if a decision-maker has no UI to another state. An analysis considering only Nash stability assumes that each decision-maker looks ahead by only one move and does not consider the potential end states that could arise if another decision-maker responds to their move. General Metarational Stability (GMR) [53], on the other hand, considers what another decision-maker could do in response to a decision-maker taking advantage of a UI. If a focal decision-maker has a UI to a second state and there exist no other decision-makers who have unilateral moves from the second state to a third state that is less preferred by the first decision-maker than it is not GMR stable [53]. GMR, then, evaluates stability of a state assuming that a decision-maker will consider possible counter moves and does not consider the preferences of other decision-makers.

Symmetrical Metarationality (SMR) [54] is similar to GMR but an initial decision-maker assumes they will have a chance to counter respond to the other decision-makers response. In Sequential Stability (SEQ) the focal decision-maker is aware of the other decision-makers' preferences and assumes that all decision-makers will only take advantage of UIs. Other solution concepts are available to capture longer time frames and different decision-maker behavior profiles [8]. In this study, the authors will focus on Nash and SEQ stability because we have no reason to believe that any decision-makers will act against their own interest simply to harm the other decision-makers [8, 50, 54].

Once stability is calculated from each decision-maker's perspective, an overall equilibrium is determined. A state is stable for a given decision-maker if that decision-maker has no unilateral improvement from that state according to a chosen stability definition. A state is an equilibrium if it is stable for all decision-makers [8, 9, 50, 54]. An equilibrium is a potential end state or solution. That a state is an equilibrium means that once reached, the state is unlikely to change unless a key component of the conflict is altered. It does not imply that decision-makers will consider the equilibrium to be a satisfactory solution.

The Decision Support Systems (DSSs) GMCR II [8] and GMCR + [55] are available to support implementation of the GMCR. The GMCR + DSS was used for this research.

5 The Teedon Pit Model

5.1 Overview of Model Development

The authors collected information related to the Teedon Pit conflict from publicly available sources. For example, news items were collected using a standard google search with keywords including, but not limited to, aggregate mine, aggregate conflict, Tiny Town, pure water, and Teedon pit. Documents were downloaded from the Township of Tiny website and from the Environmental Registry of Ontario. We reviewed key guidance documents, such as the Aggregate Resources Act, selected applicant documents posted to government and proponent websites as part of the application process, and scholarly work related to aggregate mining. From these documents, we isolated information related to the conflict components needed to build a graph model.

5.2 Decision-Makers and Their Options

In GMCR a decision-maker is defined as an entity who controls an option in the conflict at the time of analysis; Stakeholders will not always meet this criterion. Moreover, decision-makers who hold the same options and preferences can be modeled as a single decision-maker. For example, opponents to the license prefer that the expansion is not approved. One organization, the Federation of Tiny Township Shoreline Associations (FoTTSA) has engaged in two appeal processes around the Teedon Pit, demonstrating that they have the option of appealing decisions to the Ontario Land Tribunal. While researchers studying the site do not control any options, by explaining the potential benefits of their research proposal and the unique status of the site, they may significantly influence the preferences of other decision-makers. The set of decision-makers included in the model and their respective options are summarized in Table 1.

CRH Canada Group Inc., which owns Dufferin Aggregates, is the landowner and applicant at the center of this conflict. CRH is limited to two options. CRH controls the option to pursue or abandon the Teedon Pit application. While there are many steps involved in this overall application, this option can be simplified simply to apply or not apply. CRH also has the right to appeal decisions related to their application to the OLT. They can also appeal to the OLT if the Township does not respond to application materials in a timely manner [12].

The local government, represented by the Township of Tiny, controls the option to approve or reject applications under their purview related to Township plan amendments, zoning and by-laws. In January 2020, residents opposing the application were aligned with the Local Government. At that time the Township denied the CRH applications related to a water taking permit and expressed that they were "fundamentally opposed to the extraction and washing of aggregate in environmentally sensitive areas" [56]. They also formally expressed a desire to protect the area for research purposes [56, 57]. However, on May 21, 2021, the Township Council reviewed new application materials from the proponent addressing technical concerns raised in their previous notice. The Township then settled with the company thereby dropping their opposition to the application and differentiating their perspective on the conflict from the residents opposing the application (Opposition) [29, 30]. LG further distanced themselves from the Opposition perspective in 2021. At that time, a representative of the Friends of the Waverley Uplands community group requested that the Township place a moratorium on all new and expanded extraction and water taking applications in the area and that the LG enter a research partnership addressing the aquifer. LG did not act on these requests [24].

MNDMNRF is the Ontario Ministry responsible for issuing licenses and permits for aggregate mining activities. In this conflict, the MNDMNRF can issue a license for the Teedon expansion in response to an application by CRH. If the OLT directs the Minister to accept the application with added conditions, the MNDMNRF can choose to impose or remove those conditions in issuing the license [12].

Many individuals and organizations are interested in the Teedon Pit expansion application. For example, groups including the Friends of the Waverley Uplands, FoTTSA, AWARE Simcoe, Anishinaabe Kweag, and the Waterkeepers have all mobilized to protect the Alliston Aquifer from unwanted land uses [2, 22]. Their preferences are aligned with researchers who would like the aquifer to be protected while it is under study. Given that other environmental organizations may find ways to engage in appeals as FoTTSA has done, we define one decision-maker broadly as 'Opposition' and they control the option to bring an appeal to the OLT should they disagree with decisions made by the Provincial and Local governments. Opposition may control another option, escalating the conflict to the federal government by way of a petition launched seeking federal intervention through mechanisms such as the Canada Water Act [58]. The petition initiated by Opposition raises the question of federal intervention in this dispute through legislative tools such as the Canadian Water Act (CWA) [59]. It is unclear if the federal government will interpret the CWA as relevant in this case but given the unique quality of the water and its potential contributions to research, it is modeled as a possibility.

Table 1. Decision-makers and their options

Decision-maker	Option		Explanation
Canada group (CRH)	1. Apply	Yes	Apply for expansion
		No	Do not apply for expansion
	2. Bring the case to OLT[1]	Yes	Appeal decision
		No	Do not appeal decision
LG[2]	3. Approve amendment	Yes	Approve relevant applications
		No	Do not approve applications
Province	4. Approve license	Yes	Issue the license for expansion
		No	Do not issue the license
	5. MZO[3]	Yes	Issue an MZO
		No	Do not issue an MZO
	6. Approve with condition	Yes	Approve with added conditions
		No	Do not approve with conditions
Opposition	7. Bring the case to OLT	Yes	Appeal decision
		No	Do not appeal decision
	8. Petition federal government	Yes	Request federal intervention
		No	Do not request federal intervention

(*continued*)

Table 1. (*continued*)

Decision-maker	Option		Explanation
Federal government	9. Intervene	Yes	Intervene in decision
		No	Do not intervene in the conflict
OLT	10. Approval	Yes	Approves the application
		No	Rejects the application
	11. Condition	Yes	Approves application with conditions added
		No	Do not approve with conditions added

[1] Ontario Land Tribunal.
[2] Local Government.
[3] Minister's Zoning Orders.

OLT is an appeals tribunal that adjudicates conflicts related to land use planning and other related issues [https://olt.gov.on.ca, 12]. If a decision-maker refers a decision to the OLT, the OLT can then direct the MNDMNRF to accept the CRH application, to reject the CRH application, or to accept the CRH application if added conditions deemed essential by the OLT are met [12].

5.3 Feasible States and Moves in the Teedon Expansion Conflict

In a conflict with 11 options that can either be taken (Y) or not taken (N), there are 2048 mathematically possible states. However, many of those states are impossible or highly unlikely in real world conflicts, and as such, they are removed from the model. For example, we removed states where CRH did not pursue their application, but the LG approved the application. Similarly, we removed states that did not include a CRH application but included an application approval by the Province. We also removed states where the Opposition did not petition the federal government, but the federal government intervened independently. The first two examples are logically impossible while the last example is considered very unlikely. By removing a range of logically infeasible and very preferentially unlikely states in this manner, we removed 2010 infeasible states from the model.

The set of feasible states in the Teedon expansion conflict is summarized in Table 2. Decision-makers are shown to the left of the table, with the option that each decision-maker controls in the second column. Next, states are depicted column by column. A given state consists of the strategy choices of each decision-maker over their own options. In the column describing a given state, an N means that for that state, the corresponding option is not taken whereas a Y indicates that for that state the corresponding option is taken.

GMCR allows consideration of irreversible and reversible moves. Because CRH can initiate, withdraw, and reinitiate applications their option to apply is reversible. Similarly, decision-makers can withdraw from OLT hearings. All options involving approval of the application and the option for the federal government to intervene are modeled as irreversible. Options in which the application is accepted are considered irreversible.

5.4 Decision-Maker Preferences

When using GMCR+ preferences can be ranked using Option Prioritization. In this process, logical statements are used to sort the full set of states by inputting options that a given decision-maker prefers, ranked by importance. For example, based on researcher judgment we propose that CRH most prefers to have the license issued. This means that for CRH, the most important criterion for a preferred state is that Option 4 is set to 'Y'. We also suggest that they would next prefer to have the least amount of intervention or conflict. As such, states with Option 3 set to Y, and option 9 set to 'N' are preferred based on assumptions that CRH would also prefer to have no conditions added to their application and would prefer to avoid an appeals process. After inputting decision rules, 'fine tuning' by ordering states manually allows the users to ensure that rankings best reflect their interpretation of decision-makers preferences.

Table 2. Feasible states

Decision-maker	Option	States																			
		1	2	3	4	5	6	7	8	9	10	11	12	13	14	15	16	17	18	19	20
CRH	Apply	N	Y	Y	Y	Y	Y	Y	Y	Y	Y	Y	Y	Y	Y	Y	Y	Y	Y	Y	Y
	OLT	N	N	Y	N	Y	N	N	N	N	N	Y	N	Y	N	N	N	N	N	Y	Y
LG	Approve	N	N	N	Y	Y	Y	Y	N	Y	N	N	Y	Y	Y	N	Y	Y	N	Y	N
Province	Approve	N	N	N	N	N	Y	Y	Y	N	N	N	N	Y	Y	Y	Y	Y	Y	N	N
	MZO	N	N	N	N	N	N	Y	N	N	N	N	N	N	Y	N	N	Y	N	N	N
	Approve with conditions	N	N	N	N	N	N	N	N	N	N	N	N	N	N	N	N	N	N	N	N
Opposition	OLT	N	N	N	N	N	N	N	Y	N	N	N	N	N	N	Y	N	N	Y	N	N
	Petition	N	N	N	N	N	N	N	N	Y	Y	Y	Y	Y	Y	Y	Y	Y	Y	N	N
Federal	Intervene	N	N	N	N	N	N	N	N	N	N	N	N	N	N	Y	Y	Y	N	N	N
OLT	Approval	N	N	N	N	N	N	N	N	N	N	N	N	N	N	N	N	N	N	Y	Y
	Add conditions	N	N	N	N	N	N	N	N	N	N	N	N	N	N	N	N	N	N	N	N

Decision-maker	Option	States																	
		21	22	23	24	25	26	27	28	29	30	31	32	33	34	35	36	37	38
CRH	Apply	Y	Y	Y	Y	Y	Y	Y	Y	Y	Y	Y	Y	Y	Y	Y	Y	Y	Y
	OLT	N	Y	Y	N	N	Y	Y	Y	Y	N	N	Y	Y	Y	N	N	N	N
LG	Approve	Y	N	Y	Y	Y	N	Y	N	Y	Y	Y	N	Y	N	Y	Y	Y	Y
Province	Approve	Y	N	N	Y	Y	N	N	N	N	Y	N	N	N	N	N	Y	N	Y
	MZO	N	N	N	N	N	N	N	N	N	N	N	N	N	N	N	N	N	N
	Approve with conditions	N	N	N	N	N	N	N	Y	Y	N	Y	N	N	Y	Y	N	Y	N
Opposition	OLT	Y	N	N	Y	Y	N	N	N	N	Y	Y	N	N	N	N	Y	Y	Y
	Petition	N	Y	Y	Y	Y	N	N	N	N	N	Y	Y	Y	Y	Y	Y	Y	Y
Federal	Intervene	N	N	N	N	Y	N	N	N	N	N	N	N	N	N	N	N	N	Y
OLT	Approval	Y	Y	Y	Y	Y	N	N	N	N	N	N	N	N	N	N	N	N	N
	Add conditions	N	N	N	N	N	Y	Y	Y	Y	Y	Y	Y	Y	Y	Y	Y	Y	Y

We repeat this process for each decision-maker based on preferences derived from combined researcher judgment, statements made in the media, journalistic and academic accounts, previous behavior, and constraints expressed in planning documents and relevant legislation. Preference rankings for each decision-maker are displayed in Table 3. For each decision-maker the set of states is ordered such that more preferred states are listed to the left with less preferred states placed to the right in an extended pairwise fashion. States enclosed in square brackets are equally preferred. For example, in Table 3 Opposition most prefers state 1 in which CRH ceases their application. Next, they equally prefer states 2 and 9 in which LG and Province have not approved the Teedon expansion and CRH has not appealed to the OLT. Their least preferred options are 7 and 14 in which the province has issued an MZO and approved the application.

LG would most prefer that the expansion application is abandoned. LG then prefers states for which the CRH does not appeal their decision at the OLT. Next, they prefer that Opposition does not appeal a decision at OLT. We speculate that LG would like to see the application rejected by the province but would prioritize avoiding conflict. The province would most prefer that the Federal government does not intervene in their jurisdiction. The province's other preferences related to the application are unclear at this time. However, most aggregate mining licenses are approved [16], indicating a preference for issuing licenses. The province would prefer to add conditions rather than initiate an MZO which may be controversial.

Opposition most prefers that the expansion does not occur and that if it does occur, there are conditions added to protect the aquifer. We suggest that the federal government would prefer not to intervene in provincial jurisdictions. Other preferences are unknown. Finally, it is anticipated that the OLT will remain neutral until they are engaged in an appeal and hear relevant arguments.

Table 3. Decision-maker state preferences ranked from most to least preferred

Decision-maker	States ranked by preferences from most to least preferred.
CRH	6, 13, 21, 30, 8, 24, 36, 15, 7, 14, 20, [27, 29], 23, [33, 35], [11. 12], 31, 37, 19, [26, 28], 22, [2, 3, 4, 5, 9, 10, 32, 34], 1, [16, 17, 18, 25, 38]
LG	1, [2, 4,5, 9, 11], [20, 23, 27, 33], [6, 13, 16], [7, 14, 17], [31, 37], [8, 15, 18, 21, 24, 25, 30, 36, 38], [28, 29, 34, 35], [3, 10, 12, 19, 22, 26, 32]
Province	6, 13, 8, 21, 15, [24, 30, 36], [28, 29, 31, 34, 35, 37], [7, 14], [1, 2, 3, 4, 5, 9, 10, 11, 12, 19, 20, 22, 23, 26, 27, 32, 33], [16, 18, 25, 38] 17]
Opposition	1, [2,9], [3, 10], [26, 32], [19, 22], [4, 11], [5, 12], [27, 33, 20, 23], 16, 17, 18, 38, 25, [28, 34], [29, 35], [31, 37], [6, 13], [8, 15], [30, 36], [21, 24], [7, 14]
Federal	[1, 2, 3, 4, 5, 6, 7, 8, 19, 20, 21, 26, 27, 28, 29, 30, 31], [9, 10, 11, 12, 13, 14, 15, 16, 17, 18, 22, 23, 24, 25, 32, 33, 34, 35, 36, 37, 38]
OLT	[1, 2, 4, 6, 7, 9, 11, 13, 14, 16, 17, 3, 5, 8, 10, 12, 15, 18, 19, 20, 21, 22, 23, 24, 25, 26, 27, 28, 29, 30, 31, 32, 33, 34, 35, 36, 37, 38]

6 Stability Results

The Teedon Pit expansion conflict model produced 38 feasible states. In 21 of the states, the province approves the application (with or without added conditions), showing that the evaluation process provides more opportunities for approval. However, we identify 5 states in which the federal government intervenes given the unique qualities of the aquifer under threat.

Unilateral moves are summarized in Table 4, with UIs differentiated in bold type. For example, CRH has UIs from state 1 to states 2 and 3, because they have the power to initiate an application and to appeal to OLT. LG has 5 UIs largely based on their ability to avoid an OLT appeal from the CRH by approving application materials. The province has 8 UIs, showing they hold greater control in pursuing their objectives. Opposition has five UIs, but none of these UIs result in a highly preferred state in which the application is rejected. The federal government and OLT do not have any UIs as they were modeled as fairly neutral and neutral respectively. As more information surfaces regarding the federal and OLT preferences, updating this aspect of the model will be an important avenue for future research.

Stability analysis indicates which states are in equilibrium based on a chosen solution concept. As indicated in Sect. 4, if a given state is stable for all decision-makers, then it is an equilibrium. If, on the other hand, it is unstable for any decision-maker, then the conflict will not remain at that state, and it is not an equilibrium.

Consider, for example, state 4. CRH, Province, and Opposition each have a UM from state 4. However, for CRH and Opposition, the state they can reach from state 4 is not more preferred than state 4 is from their own perspective. Thus, for CRH and Opposition, there are no UIs from state 4 and state 4 is stable. The Province on the other hand, can move the conflict from state 4 to state 6 by controlling their own option to approve the license. Since state 6 is more preferred than state 4 from the perspective of the Province, the Province is motivated to move the conflict from state 4 to state 6. As such, state 4 is not stable for all decision-makers and is not an equilibrium when considering Nash stability. However, the Province may consider the consequences of taking advantage of their UI from state 4 to state 6. If another decision-maker has a UI from state 6 that moves the conflict to a new state that is less preferred than initial state 4, then the Province would prefer to avoid ending up at a less preferred state and will remain at state 4. Opposition has 3 UMs from state 5, but none of them are improvements for Opposition, so they are not UIs. Accordingly, by SEQ stability, state 4 is still considered unstable because the Province benefits from moving to state 6.

Considering GMR stability, state 4 would remain unstable because the Opposition UMs from state 6 bring the conflict to states that are more preferred than state 4 from the perspective of the Province. As such, the Opposition UMs do not deter the Province from using their own UI to move the conflict from state 4 to state 6.

Table 4. Unilateral moves in the tiny town conflict. UIs are shown in bold type

Initial State	CRH	LG	Province	Opposition	Federal	OLT
1	2, 3					
2	1, 3	**4**	**7**	9		
3	1, 2	**5**		10		19, 26
4	5		**6**	11		
5	4			12		20, 27
6				8, 13, 15		
7				14		
8				6, 13, 15		21, 30
9	10	**11**	**14**	2		
10	9	12		3		22, 32
11	12		**13**	4		
12	11			5		23, 33
13				6, 8, 15	16	
14				7	17	
15				6, 8, 13	18	24, 36
16				18		
17						
18				16		25, 38
19		**20**		22		
20				23		
21				24		
22		**23**		19		
23				20		
24				21	25	
25						
26		**27**	**28**	32		
27			**29**	33		
28		**29**		34		
29				35		
30				36		
31				37		
32		**33**	**34**	26		
33			**35**	27		
34			**35**	28		
35				29		
36				30	38	
37				31		
38						

In this analysis, Nash and SEQ stability are considered and GMR is not. This is because there are no indications that any decision-makers are motivated to disadvantage or sanction other decision-makers at a cost to themselves. Twenty-three states are equilibria, including states [5–7, 10, 12–14, 16, 17, 20, 21, 23–25, 28–31, 34–38]. The equilibria set includes 18 states in which the application is approved by the province, including with and without conditions, reinforcing again the limited number of scenarios in which the expansion is completely rejected.

7   Discussion and Concluding Thoughts

The Teedon Pit expansion application is modeled using the GMCR, capturing key decision-makers and the opportunities they have to direct the evolution of this conflict. The model illustrates that there are more opportunities for the application to be accepted than not. However, the unique purity of the water may lead to the uncommon opportunity for federal intervention in determining the future of the Alliston aquifer. Because Provinces have the authority to grant or refuse licenses and to issue MZOs, changing the venue where aggregate mining decisions are made in an effort to improve possible outcomes, a strategy known as venue shopping [60], may not typically be a viable option in these contexts [61]. This option is an interesting strategy choice by the Opposition and distinguishes the Tiny Town conflict from other aggregate mining disputes [61].

The limited number of UIs controlled by the proponent in this model should be interpreted with care. While this suggests that the applicant has a limited role in the conflict outside of their ability to choose to submit, withdraw, and resubmit application material and to appeal decisions or delays related to the application, such an inference is misleading. Their land ownership status and control over the timeline of applications enables proponents to control the information under consideration by the other decision-makers [16]. This model provides a foundation on which to conduct a comprehensive case study of this conflict that could further investigate the nuanced drivers of conflict evolution in this case.

Land use conflicts are a special type of dispute where competing interests, democratic ideals, and institutional powers are negotiated, manifested, and propagated. As such, they need to be documented, analyzed, and evaluated for their provincial and national consequences. The GMCR provides one way to document and investigate such events. The limited informational demands regarding decision-maker preferences are particularly well suited to these types of controversial land use conflicts where decision-makers may be reluctant to openly state their preferences in the conflict.

References

1. Mittelstaedt, M.: The battle over the world's purest water. The globe and mail. 4 May 2009 (2009). https://www.theglobeandmail.com/news/national/the-battle-over-the-worlds-purest-water/article4193063/. Accessed 04 Apr 2022.
2. Mendler, A.: Newsmakers of the year: the pristine water of tiny township. Simcoe.com. 27 Dec 2018 (2018). https://www.simcoe.com/news-story/9083810-newsmakers-of-the-year-the-pristine-water-of-tiny-township/. Accessed 04 Apr 2022

3. Mendler, A.: 'We have to protect this': tiny township 'waterkeepers' taking gravel pit concerns to Ontario legislature. The Star. 14 Mar 2022 (2022). https://www.simcoe.com/news-story/10586704--we-have-to-protect-this-tiny-township-waterkeepers-taking-gravel-pit-concerns-to-ontario-legislature/. Accessed 04 Apr 2022

4. Mendler, A.: 'It's important to all Canadians': Petition calls on federal government to protect Tiny Township aquifer. The Mirror. 19 Mar 2022 (2022). https://www.thestar.com/local-midland/news/2022/03/19/it-s-important-to-all-canadians-petition-calls-on-federal-government-to-protect-tiny-township-aquifer.html. Accessed 30 Mar 2022

5. Cecco, L.: The Canadian town of Tiny has the world's purest water. A gravel mining operation could ruin it. The Guardian, Nov 2021 (2021). https://www.theguardian.com/environment/2021/nov/25/tiny-town-purest-water-georgian-bay-ontario-gravel-mining.25. Accessed 04 Apr 2022

6. Ioanna, R., Witmer, B.: Residents fear effects of increasing quarry activity on Elmvale groundwater, believed to be cleanest in world. CBC News. 27 Nov 2021 (2021). https://www.cbc.ca/news/canada/elmvale-ground-water-quarry-1.6199158. Accessed 04 Apr 2022

7. Marc Kilgour, D., Hipel, K.W., Fang, L.: The graph model for conflicts. Automatica **23**(1), 41–55 (1987). https://doi.org/10.1016/0005-1098(87)90117-8

8. Fang, L., Hipel, K.W., Kilgour, D.M.: Interactive Decision Making: The Graph Model for Conflict Resolution. Wiley, Hoboken (1993)

9. Xu, H., Hipel, K.W., Kilgour, D.M., Fang, L.: Conflict Resolution Using the Graph Model: Strategic Interactions in Competition and Cooperation. Springer, Cham (2018). https://doi.org/10.1007/978-3-319-77670-5

10. Hipel, K.W., Fang, L., Kilgour, D.M.: The graph model for conflict resolution: reflections on three decades of development. Group Decis. Negot. **29**(1), 11–60 (2019). https://doi.org/10.1007/s10726-019-09648-z

11. Hipel, K.W., Fang, L.: The graph model for conflict resolution and decision support. IEEE Trans. Syst. Man Cybern. Syst. **51**(1), 131–141 (2021). https://doi.org/10.1109/TSMC.2020.3041462

12. Aggregate Resources Act.: 2019. Pub. L. No. R.S.O. 1990, Chapter A.8 (2019). https://www.ontario.ca/laws/statute/90a08. Accessed 25 Feb 2021

13. Drew, L.J., Langer, W.H., Sachs, J.S.: Environmentalism and natural aggregate mining. Nat. Resour. Res. **11**(1), 19–28 (2002). https://doi.org/10.1023/A:1014283519471

14. Sandberg, L.A.: Leave the sand in the land, let the stone alone: pits, quarries and climate change. ACME Int. J. Crit. Geog. **12**(1), 65–87 (2013)

15. Wagner, E.V.: Law's rurality: land use law and the shaping of people–place relations in rural Ontario. J. Rural. Stud. **47**, 311–325 (2016). https://doi.org/10.1016/j.jrur-stud.2016.01.006

16. Wagner, E.V.: The work of ownership: shaping contestation in ontario's aggregate extraction disputes. In: Bruun, M.H., Cockburn, J.L., Risager, B.S., Thorup, M. (eds.) Contested property claims. What disagreement tells us about ownership, Chapter 7. Taylor and Francis. (2018)

17. Constitution Act. 1867.: (UK), 30 & 31 Vict, c 3, reprinted in RSC 1985, App II, No 5. Accessed 25 Feb 2021. https://www.canlii.org/en/ca/laws/stat/30%13-31-vict-c-3/97547/30%13-31-vict-c-3.html

18. Provincial policy statement 2020.: Approved by the lieutenant governor in council, Order in council no. 229/2020 (2020). https://files.ontario.ca/mmah-provincial-policy-statement-2020-accessible-final-en-2020-02-14.pdf. Accessed 25 Feb 2021

19. Planning Act, R.S.O. 1990 (2021). https://www.ontario.ca/laws/statute/90p13#BK8. Accessed 16 Mar 2022

20. Statistics Canada 2017.: Tiny, TP [Census subdivision], Ontario and Saskatchewan [Province] (table). Census Profile. 2016 Census. Statistics Canada Catalogue no. 98-316-X2016001. Ottawa. Released 29 Nov 2017. https://www12.statcan.gc.ca/census-recensement/2016/dp-pd/prof/index.cfm?Lang=E. Accessed 31 Mar 2022

21. Township of Tiny. (n.d.).: Living in Tiny. Retrieved 28 Mar 2022. https://www.tiny.ca/tow nship-hall/about-tiny. Township of Tiny 2020. https://www.tiny.ca/Shared%20Documents/ Teedon%20Pit%20Expansion%20Aggregate%20Resources%20Act%20Application/Tow nship%20Letter.pdf. Accessed 2 Apr 2022

22. Shahid, M.: 'David and Goliath' clash looming over Tiny's Teedon Pit extension. OrilliaMat ters.com. 23 Jan 2021 (2021). https://www.midlandtoday.ca/local-news/david-and-goliath-clash-looming-over-tinys-teedon-pit-extension-3285324. Accessed 04 Apr 2022

23. Shotyk, W., Krachler, M., Aeschbach-Hertig, W., Hillier, S., Zheng, J.: Trace elements in recent groundwater of an artesian flow system and comparison with snow: enrichments, depletions, and chemical evolution of the water. J. Environ. Monit. **12**, 208–217 (2010)

24. Howard, D.: Tiny exploring whether it's taxing aggregate companies 'correctly'. Midland Today. 13 Jan 2022 (2022). https://www.midlandtoday.ca/local-news/tiny-exploring-whether-its-taxing-aggregate-companies-correctly-4951430. Accessed 04 Apr 2022

25. Mendler, A.: Expanded tiny township gravel pit to impact environment minimally: com pany. Simcoe.com. 6 Mar 2019 (2019). https://www.simcoe.com/news-story/9208653-exp anded-tiny-township-gravel-pit-to-impact-environment-minimally-company/. Accessed 04 Apr 2022

26. Bacher, J.: Alliston aquifer threatened again by gravel washing. Sierra Club Foundation. 20 Jan 2021 (2021). https://www.sierraclub.ca/en/ontario-chapter/2021-01-20/alliston-aquifer-threatened-again-gravel-washing

27. MHBC (MacNaughton, Hermson Britton Clarkson Planning Limited).: Aggregate Resources Act Summary Statement. January 2019 (2019). https://www.tiny.ca/Shared%20Documents/ Teedon%20Pit%20Expansion%20Aggregate%20Resources%20Act%20Application/Agg regate%20Resources%20Act%20Summary%20Statement-Teedon%20Pit%20Extension-09Jan2019.pdf. Accessed 30 Mar 2022

28. Government of Ontario.: Environmental Registry of Ontario (ERO) (2021). https://ero.ontari o.ca/

29. Esemag.com.: Tiny Township groups stay in fight to protect pure water against gravel mining. 6 Dec 2021 (2021). https://esemag.com/water/tiny-township-fight-protect-water-against-gra vel-mining/. Accessed 04 Apr 2022

30. Harries, K.: Purest water: Tiny council votes 3–2 to settle with aggregate company. Aware Sim coe. 22 Jun 2021 (2021). https://aware-simcoe.ca/2021/06/purest-water-tiny-council-votes-3-2-to-settle-with-aggregate-company/. Accessed 04 Apr 2022

31. Hipel, K.W., Kilgour, D.M., Fang, L., Peng, X.: The decision support system GMCR II in negotiations over groundwater contamination. In: Proceedings of the 1999 International Conference on Systems, Man, and Cybernetics, vol. 5, pp. 942–948. IEEE, Tokyo (1999)

32. Hipel, K.W., Kilgour, D.M., Fang, L., Peng, X.: Applying the decision support system GMCR II to negotiation over water. In: Negotiation over water, ed. U. Shamir, 50–70. The International Hydrological Programme, Technical Document in Hydrology No. 53. Paris: United Nations Educational, Science and Cultural Organization (2001)

33. Akbari, A., Mirnasl, N., Hipel, K.W.: Will peaceful waters flow again? A game-theoretic insight into a tripartite environmental conflict in the Middle East. Environ. Manage. **67**(4), 667–681 (2021). https://doi.org/10.1007/s00267-021-01429-2

34. Madani, K., Hipel, K.W.: Non-cooperative stability definitions for strategic analysis of generic water resources conflicts. Water Resour. Manage **25**(8), 1949–1977 (2011). https://doi.org/ 10.1007/s11269-011-9783-4

35. Chu, Y., Hipel, K.W., Fang, L., Wang, H.: Systems methodology for resolving water conflicts: the Zhanghe river water conflict in China. Int. J. Water Resour. Dev. **31**(1), 106–119 (2015). https://doi.org/10.1080/07900627.2014.933096

36. Philpot, S.L., Hipel, K.W., Johnson, P.A.: Strategic analysis of a water rights conflict in the south-western United States. J. Environ. Manage. **180**, 247–256 (2016). https://doi.org/10.1016/j.jenvman.2016.05.027
37. Garcia, A., Hipel, K.W., Obeidi. A.: Water pricing conflict in British Columbia. Hydrol. Res. Lett. **11**(4), 194–200 (2017). https://doi.org/10.3178/hrl.11.194
38. Noakes, D.J., Fang, L., Hipel, K.W., Kilgour, D.M.: An examination of the salmon aquaculture conflict in British Columbia using the graph model for conflict resolution. Fish. Manage. Ecol. **10**(3), 123–137 (2003). https://doi.org/10.1046/j.1365-2400.2003.00336.x
39. Noakes, D.J., Fang, L., Hipel, K.W., Kilgour, D.M.: The pacific salmon treaty: a century of debate and an uncertain future. Group Decis. Negot. **14**(6), 501–522 (2005). https://doi.org/10.1007/s10726-005-9005-7
40. Hu, K., Hipel, K.W., Fang, L.: A conflict model for the international hazardous waste disposal dispute. J. Hazard. Mater. **172**(1), 138–146 (2009). https://doi.org/10.1016/j.jhazmat.2009.06.153
41. Bernath Walker, S., Boutilier, T., Hipel, K.W.: Systems management study of a private brown-field renovation. J. Urban Plann. Dev. **136**(3), 249–260 (2010). https://doi.org/10.1061/(ASCE)0733-9488(2010)136:3(249)
42. Hipel, K.W., Hegazy, T., Yousefi, S.: Combined strategic and tactical negotiation methodology for resolving complex brownfield conflicts. Pesquisa Operacional **30**(2), 281–304 (2010). https://doi.org/10.1590/S0101-74382010000200003
43. Yousefi, S., Hipel, K.W., Hegazy, T.: Considering attitudes in strategic negotiation over brown-field disputes. J. Leg. Aff. Disput. Resolut. Eng. Constr. **2**(4), 240–310 (2010). https://doi.org/10.1061/(ASCE)LA.1943-4170.0000034
44. Yousefi, S., Hipel, K.W., Hegazy, T.: Optimum compromise among environmental dispute issues using attitude-based negotiation. Can. J. Civ. Eng. **38**(2), 184–190 (2011). https://doi.org/10.1139/L10-125
45. Hipel, K.W., Walker, S.B.: Brownfield redevelopment. In: Craig, R. (ed.) Ecosystem management and sustainability, pp. 44–48. Barrington (2012)
46. Philpot, S.L., Johnson, P.A., Hipel, K.W.: Analysis of a brownfield management conflict in Canada. Hydrol. Res. Lett. **11**(3), 141–148 (2017). https://doi.org/10.3178/hrl.11.141
47. Xiao, Y.K., Hipel, K.W., Fang. L.: Strategic investigation of the Jackpine mine expansion dispute in the Alberta oil sands. Int. J. Decis. Support Syst. Technol. **71**(1), 50–62 (2015)
48. Garcia, A., Obeidi, A., Hipel, K.W.: Two methodological perspectives on the energy East Pipeline conflict. Energy Policy **91**, 397–409 (2016). https://doi.org/10.1016/j.enpol.2016.01.033
49. Fang, L., Hipel, K.W., Kilgour, D.M., Peng, X.: A decision support system for interactive decision making—Part I: model formulation. IEEE Trans. Syst. Man Cybern. Part C (Appl. Rev.) **33**(1), 42–55. (2003). https://doi.org/10.1109/tsmcc.2003.809361
50. Fang, L., Hipel, K.W., Kilgour, D.M., Peng, X.: A decision support system for interactive decision making—Part II: analysis and output interpretation. IEEE Trans. Syst. Man Cybern. Part C (Appl. Rev.) **33**(1), 56–66. (2003). https://doi.org/10.1109/tsmcc.2003.809360
51. Nash, J.: Equilibrium points in n-person games. In: Proceedings of the National Academy of Sciences of the U.S.A, vol. 36, pp. 48–49 (1950)
52. Nash, J.: Noncooperative games. Ann. Math. **54**(2), 286–295 (1951). https://doi.org/10.2307/1969529
53. Howard, N.: Paradoxes of Rationality. MIT Press, Cambridge (1971)
54. Fraser, N., Hipel, K.W.: Conflict Analysis Models and Resolutions. Elsevier Science, New York (1984)
55. Kinsara, R.A., Petersons, O., Hipel, K.W., Kilgour, D.M.: Advanced decision support for the graph model for conflict resolution. J. Decis. Syst. **24**(2), 117–145 (2015). https://doi.org/10.1080/12460125.2015.1046682

56. Township of Tiny.: Confidential Report. Motion # 249. Meeting Date: 21 May 2021 (2021). https://www.tiny.ca/Shared%20Documents/Teedon%20Pit%20Expansion%20Aggregate%20Resources%20Act%20Application/Council%20Motion%20249%20and%20Staff%20Report%20%20PD-027-21.pdf. Accessed 04 Apr 2022
57. Mendler, A.: Tiny Township 'fundamentally opposed' to aggregate extraction in 'environmentally sensitive areas' Simcoe.com. 4 Feb 2020 (2020). https://www.simcoe.com/news-story/9839678-tiny-township-fundamentally-opposed-to-aggregate-extraction-in-environmentally-sensitive-areas-/. Accessed 04 Apr 2022
58. Philips, A.: Petition seeks federal help to protect 'purest water in the world'. OrilliaMatters.com 3 Mar 2022 (2022). https://www.midlandtoday.ca/local-news/petition-seeks-federal-help-to-protect-regions-purest-water-in-the-world-5122301. Accessed 04 Apr 2022
59. CWA. 1985. Canada Water Act, RSC 1985, c C-11. https://canlii.ca/t/527q3. Retrieved on 05 Apr 2022. Accessed 03 Apr 2022
60. Pralle, S.B.: Venue shopping, political strategy, and policy change: the internationalization of Canadian forest advocacy. J. Publ. Policy **23**(3), 233–260 (2003)
61. Philpot, S., Hipel, K.W.: Investigating an aggregate mine proposal using the Graph Model for Conflict Resolution. Ann. Am. Assoc. Geogr. 1–21 (2022). https://doi.org/10.1080/24694452.2021.1994850

Cauvery River: Path Dependence and Feedback in Water Sharing Conflicts

Ajar Sharma[1]($\boxtimes$) (iD), Keith W. Hipel[1] (iD), and Vanessa Schweizer[2] (iD)

[1] Systems Design Engineering Department, Faculty of Engineering, University of Waterloo, Waterloo, ON N2L 3G1, Canada
ajar.sharma@uwaterloo.ca

[2] Department of Knowledge Integration, Faculty of Environment, University of Waterloo, Waterloo, ON N2L 3G1, Canada

Abstract. The concepts of Path Dependency and Leverage Points were used to better understand the complex Cauvery River dispute. Pierson's theories are used to examine and analyze the historical events of the system and potential feedback loops are identified and explained. Meadows' leverage points about this conflict are described in brief, and the reasoning behind their applicability is also explained. The approach identifies the major positive feedback loops in the system and reveals a causal loop diagram.

Keywords: Conflict resolution · Path dependencies · Positive feedback · Causal loop diagram · Cauvery river dispute

1 Introduction and History

The Cauvery River is the fourth largest river in southern India. It originates in the province of Karnataka near Talakaveri and it flows through the province of Tamil Nadu (TN) where it reaches the Bay of Bengal. The 800-km river drains approximately 81,000 square kilometers [1–3]. The Cauvery River has been responsible for sustaining these two provinces for decades. Over the last 130 years, this basin has been constantly dealing with conflict over its water sharing. India was a British colony and gained independence in the year 1947. This conflict first surfaced in the year 1890 when the princely province of Mysore (roughly present-day Karnataka) wanted to build a dam on the river and the province of Madras (roughly present-day Tamil Nadu) opposed. The conflict was resolved for half a century when the two provinces signed a 50-year treaty in 1924. In 1970, four years before the previous treaty was supposed to expire, the government of India initiated a program to reach a new treaty. However, none of the two provinces could agree on the plans. After much deliberation, the Cauvery Water Disputes Tribunal (CWDT) was established in 1990. To assist the tribunal a Cauvery River Authority (CRA) was established in 1997. In 2000, the Cauvery Monitoring Committee (CMC) was added to further assist the existing CWDT and the CRA. In 2018, a Cauvery Water Management Authority was established to execute the previously passed orders by the CWDT. Since 1990, the tribunal has passed three major decisions and all of them have

D. C. Morais and L. Fang (Eds.): GDN 2022, LNBIP 454, pp. 91–101, 2022.
https://doi.org/10.1007/978-3-031-07996-2_7

been highly contested by the provinces. The objective of this research is to analyze the conflict by examining the historical key decision points in the timeline of this conflict which may have caused irreversible damage to the conflict resolution. The decision-makers were not aware that they were causing this damage because they couldn't predict them. These seemingly irreversible, and unpredictable decisions can induce a problem-solving inefficiency in the system. Over the years, the decision-makers who were made responsible for resolving the conflict, unfortunately, used the framework laid down by the previous decision-makers. These events assimilate into the decision-making framework and render the system inefficient. This is called Path Dependence. These frameworks, over time, settle themselves into self-reinforcing (positive) feedback loops and it becomes nearly impossible to break these feedback loops. This research uses a Causal Loop Diagram methodology to analyze the inherent feedback loops that are hidden. This research presents the analysis of these path dependencies and exposes the feedback loops in the Cauvery River system.

2 Methods and Results

Over the last hundred years, the Cauvery River conflict has evolved from being a classic water-sharing conflict to a climate change-affected water-sharing conflict. The decisions taken by the administrators over these years have consistently displayed a myopic outlook on resolving the conflict. These short-sighted decisions, unknowingly, created an inefficient framework for resolving the conflict. The new institutions, committees, tribunals, etc. that were established to resolve the conflict used the previous frameworks as stepping-stones. Therefore, the initial mistakes intertwined themselves in the resolution methodologies and were very difficult to overcome or even identify. This created a path dependence in the system [4–6]. The following paragraphs identify the path-dependent problems in the Cauvery River.

The central variable in the conflict is the water availability in the Cauvery River. As mentioned earlier, there are two major provinces, Tamil Nadu (downstream) and Karnataka (upstream), interested in extracting as much water as possible. The water in the river was divided proportionally based on the Tribunal's last decision regarding the conflict. Tamil Nadu uses the water predominantly for agricultural purposes, and Karnataka mainly uses the water for industrial and domestic purposes in its capital city of Bangalore. Climate change has caused major changes in the hydrological cycles in the region. The conflict has been exacerbated over the last three decades primarily because of faltering monsoons [7].

Globally, climate change has been a major discussion point in water policy circles over the last couple of decades. All the member countries of the UN signed on to the Millennium Development Goals (MDGs) in the year 2000 with one of the eight goals being ensuring environmental stability [8]. The UN realized the need to act on a holistic spectrum to be better prepared for the future. Before the MDGs, environmental protection was not that visible in policy discussions. A change in ideological thinking of the UN representatives pushed this agenda. Due to globalization, countries became more and more connected causing a further increase in the exchange of knowledge. India, as a country, also adopted certain aspects of these goals. An evaluation of its goals is represented in the figure below. A part of goal number 7 was to stop unsustainable exploitation of its natural resources and to develop water management strategies. Unfortunately, India [9] did not include that goal in its version (Fig. 1).

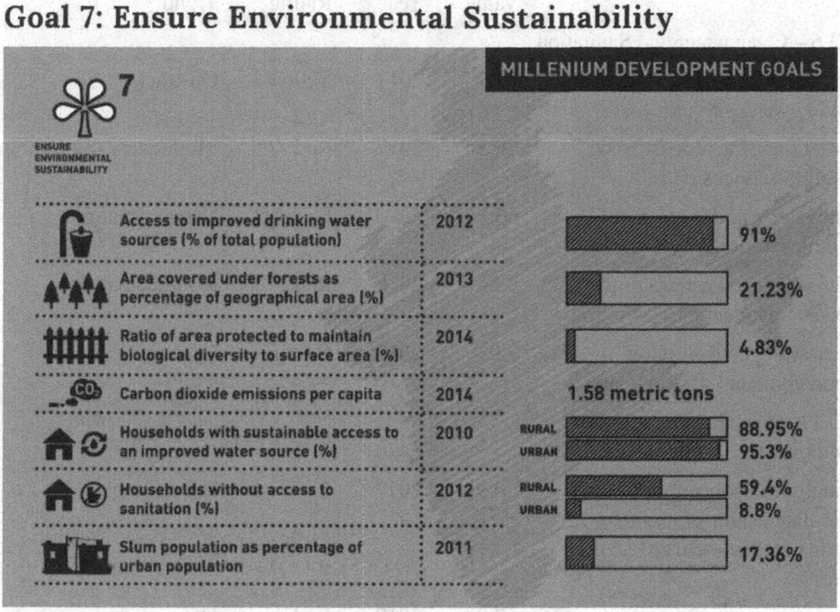

Goal 7: Ensure Environmental Sustainability

Fig. 1. Millennium development goals: India's goal 7 progress report. (Source: [9])

Of course, the MDGs were replaced by the Sustainable Development Goals (SDGs) in 2015 [10]. The number of goals was increased to 17 with much more detailed indicators to help countries make meaningful change [11]. For the Cauvery River conflict, the authors focus majorly on Goals 6 (Clean Water and Sanitation), 11 (Sustainable Cities and Communities), and 16 (Peace, Justice and Strong Institutions). Table 1 shows the performance of the country for Goals 6, 11, and 16. The table shows the indicators selected by India in the first column, its value in the second column, the year in which the indicator was achieved in the third column, the rating in the fourth column, and the trend in the fifth column for all three goals. The fourth column depicts the progress of the indicators where, red means "major challenges", orange means "significant challenges", yellow means "challenges remain", and green means "SDG achieved".

The red (major challenges), orange (significant challenges), yellow (challenges remain), and green (SDG achieved) points in Table 1 show the progress of the indicators. The fourth column shows the current trends of the indicators. Unfortunately, the more important indicators were not selected by India for any of the above indicators. One of the most important indicators for a country like India under Goal 6 is the implementation of integrated water resources management (indicator number 6.5) [11]. The Cauvery River Basin needs a proper integrated water resources management approach to solve the transboundary conflict. However, this indicator is not included in India's goals.

Table 1. Performance by indicators for Goals 6, 11, and 16 (modified from [12]).

India - Performance by indicator

	Value	Year	Rating	Trend
SDG 6 – Clean water and Sanitation				
Population using at least basic drinking water services (%)	92.7	2017	Yellow	On-track
Population using at least basic sanitation services (%)	59.5	2017	Red	Moderately improving
Freshwater withdrawal (% of available freshwater resources)	66.5	2010	Orange	No information
Anthropogenic wastewater that receives treatment (%)	2.2	2018	Red	No information
Scarce water consumption embodied in imports (m^3/capita)	2.9	2013	Green	On-track
SDG 11 – Sustainable cities and Communities				
Annual mean concentration of particulate matter of less than 2.5 microns in diameter (PM2.5) ($\mu g/m^3$)	90.9	2017	Red	Decreasing
Access to improved water source, piped (% of urban population)	67.9	2017	Red	Decreasing
Satisfaction with public transport (%)	71.9	2018	Yellow	On-track
SDG 16 – Peace, Justice and Strong institutions				
Homicides (per 100,000 population)	3.2	2016	Orange	On-track
Unsentenced detainees (% of prison population)	67.7	2018	Red	Decreasing
Percentage of population who feel safe walking alone at night in the city or area where they live (%)	69.3	2018	Yellow	On-track
Property rights (worst 1–7 best)	4.4	2019	Yellow	No information
Birth registrations with civil authority (% of children under age 5)	79.7	2018	Orange	No information
Corruption perception index (worst 0–100 best)	41	2019	Orange	Moderately Improving

(*continued*)

Table 1. (*continued*)

India - Performance by indicator				
Children involved in child labor (% of population aged 5 to 14)	11.8	2016	Red	No information
Exports of major conventional weapons (TIV constant million USD per 100,000 population)	0	2019	Green	No information
Press freedom index (best 0–100 worst)	45.7	2019	Orange	Decreasing

Similarly, within Goal 16 there was a provision to include the possibility of strengthening its institutions along with capacity building (16. 5), and substantially reducing corruption and bribery in all forms (16. a) [12] For a developing country like India, one must have robust and resilient administrative institutions to combat the complex nature of its problems. However, these indicators were not included in the goals. Over the years, there have been a few social unrest events in the Cauvery River Basin due to the politics over the sharing of water [3].

Locally, the climate action plans for the provinces of Tamil Nadu and Karnataka were published by provinces in the year 2011 and 2013, respectively [13–15]. These reports talk about the need to increase the efficiencies in the distribution of water along with efficiencies in agricultural practices. Because of faltering rainfall, the available water in the river has reached critical limits. About a century ago in 1924, when the first pact was signed between Mysore (Karnataka) and Madras (Tamil Nadu) for 50 years, a potential increase in the population of the two provinces was not taken into consideration. Although then climate change was not known or was not considered to be that important, a certain measure for geographical change due to population increase should have been part of the pact. Over the last three decades, the population of both provinces has increased heavily. The population of Karnataka, for example, has increased exponentially from 4 million in 1990 to 13 million in 2022 [16, 17].

The area of irrigation has not increased much since the late 1980s in Tamil Nadu; however, inconsistent rainfall has hampered the agricultural industry. The independence of the country in the year 1947 also played a considerable hand in the conflict due to the redrawing of the geographical boundaries. In 1970, when the Cauvery Fact-finding Committee was established under an archaic River Boards Act of 1956 [18], the law had no provision for including the provinces in the discussion and was based completely based on water resource availability. However, it was the institutional short-sightedness on the part of the federal government and the province governments when the pact signed in 1974 was broken by Tami Nadu in 1976. In the year 1986, this act was modified which led to the establishment of the 'Cauvery River Water Dispute Tribunal (CWDT)' in the year 1990. The tribunal gave its most notable decision in the year 2000, five years after the first indications of irregular rainfall. This decision included the methodology to be followed during distress years. Instead of placing a contingency plan in place, it was simply stated that the provinces should share the distress and use water judiciously. Still no provision of climate change and its effects, and no mention of the need for proactive measures to solve the imminent threat of scarcity. Even with the existing understanding of climate change in academia at the time, the inherent path dependence prevented the inclusion of climate change mitigation provisions in the tribunal's orders.

The institutions established to resolve the conflict have remained largely ineffective because the basis of their decision-making is the frameworks already established. In 1905, the then federal government tried to mediate between the provinces of Mysore (Karnataka) and Madras (Tamil Nadu) over a conflict in the region. The conflict at that time was about the construction of a dam by Mysore, which ultimately led to the present-day water-sharing conflict. At that time, Mysore refused to follow the directives of the government and went ahead with its plans. No action was taken against Mysore, and this established the first instance of a province government disobeying the federal government's orders and getting away with them. In 1974, after the completion of the 50-year pact, the two provinces of Karnataka and Tamil Nadu signed another pact, but a couple of years later in 1976, Tamil Nadu backed out of the pact. The pact was signed and still, Tamil Nadu did not adhere to its contents, and the federal government took no strict actions, rather it went on to unsuccessfully renegotiate the pact for the next 15 years. In 1990, after the CWDT was established, the first order it gave out was to nullify any provision for finding the culprit in breaking the pacts. It dictated that it was futile to get into such practices. Due to no stricter actions against the parties who breach the contract being taken, in the years to come, either of the provinces refused to adhere to the agreements. Majorly it was the province of Karnataka, as it is the upstream region. Karl Mannheim in his essay on generations talked about the importance of early impressions of a situation on our future thinking about a problem. Early impressions and frameworks tend to coalesce into a natural view of the world [19]. In the Cauvery Conflict, the early perception of an abundance of water in the region and non-belief in climate change, and the later experiences seem to have changed the view of this dispute over the years. Therefore, the clarification of specific mechanisms that generate path dependence in the system is the key to hypothesizing about the sources of social stability and change [4].

Two major path-dependent processes mentioned above are the non-inclusion of climate change effects policy and institutional ineffectiveness in regulating the provinces. Over the years, they have assimilated themselves in the process from which it is difficult to deviate. The institutions find it easier to work around the established format of dispute resolution, with the same exercise being carried out every time there is a need to mediate.

The path dependency establishes itself in the system by developing feedback loops, which are difficult to break free from. There are two types of feedback in a system, positive, shown with a ' +' sign, and negative shown with a '- 'sign. Positive feedback loops are self-reinforcing, and negative feedback loops are self-correcting [4]. Figure 2 is the final product of this research. In the following paragraphs, the process of the creation of the causal loops is explained.

Climate change (center-top of Fig. 2) negatively affects the rainfall in both provinces, which in turn affects the various boxes depicted in Fig. 2. There are self-reinforcing loops that would reinforce a behaviour if not stopped or checked. Rainfall increases cause water availability to increase in both provinces. Water availability (shown in Fig. 2) is connected to water demand, which is affected positively as well. Water demand in Tamil Nadu is affected by the water availability from the rainfall, as well as the upstream availability released by Karnataka [13, 14]. The water availability mentioned here includes the water available in the Cauvery River, and the water available due to rainfall over the agricultural fields. Another factor is flooding due to flash rains. However, its frequency of occurrence is less than that of inconsistency in rainfall. Since most of the water utilized in the region is for industrial and agricultural purposes, more water utilization causes more revenue generation. Agricultural export is the biggest contributor to Tamil Nadu's economy [15], and Bangalore is the Silicon Valley of India with billions of Rupees invested in the region's high-tech industry [17].

An increase in water supply efficiency will increase water utilization/demand. This is defined as Jevons' Paradox [20]; the increase in efficiency of utilizing a natural resource does not necessarily lead to less consumption of that resource [21]. This water supply efficiency depends upon the water demand in the respective province and therefore positively affects the revenue generation of the province. Institutions included in the Institutional Effectiveness are the Supreme Court of India, the CWDT, and the provincial governments. If the revenue generation of the province is high, the institutional effectiveness of the province is high because it can allocate resources appropriately and generate profits so that they can be utilized for the welfare of its province. Higher institutional effectiveness will also cause an increase in water distribution efficiency. If the provincial government and the federal government are allocating resources for the infrastructure development of the province and conserving water, the distribution efficiency will increase. It is currently around 50% in Bangalore. The agricultural efficiency of water use in Tamil Nadu is as low as 40% [22]. Recently, the government has taken steps in the right direction, however, there is still a long way to go [23].

The major self-correcting loop shown in Fig. 2 via social unrest has a negative connotation to it. In a democratic country like India, public dissent is generally displayed during the federal and provincial elections. However, when the citizens feel that they are being mistreated by their government, the dissent can manifest itself in other methodologies. The consequences of any actions taken depend upon the actions of others. Public dissent usually starts as a non-violent protest with candlelight marches and demonstrations in front of the provincial minister's offices. Often, however, these protests turn violent. The Cauvery River conflict has seen many such riots [3, 24] during the last two decades. Unfortunately, this quickly transforms into a divisive debate where people suddenly become aware of the well-being of "their people" and want to cause harm to others. This situation is categorized as "Social Unrest" (center of Fig. 2) in the causal loop diagram. This has caused major public and personal losses to the provinces [25]. In the year 2016, the province of Karnataka lost 25000 crores of rupees in property damages, which is close to 3.6 billion dollars. If there is more rainfall, there is less rioting as people are generally rioting in the region due to lack of water. If the water availability is higher then also the rioting is decreased. The target line from Social Unrest towards openness to negotiating with TN has a negative relationship because the higher the anti-other sentiments, the lower the willingness to have a dialogue. The willingness to negotiate (bottom of Fig. 2) with Tamil Nadu (TN) is Karnataka's prerogative, as they are upstream. Willingness to negotiate forms one of the most important aspects of this causal loop diagram as well as the conflict. There have been some signs of improvement in the government's work concerning this situation. For example, in Tamil Nadu, after the last riots (2016), the opposition party in the province started campaigning on an ineffective incumbent government platform alleging that an ineffective government may give rise to an increase in instances of corruption. In response to the opposition political party, the incumbent provincial political party staged a protest against the federal government to transfer the blame. Since 2016, people have become more aware of how governments are manipulating and affecting them. In the context of this conflict, having an effective government is extremely important.

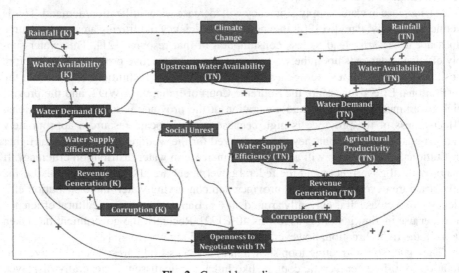

Fig. 2. Causal loop diagram

From the above analysis of the conflict and with the influence diagrams in play in the legal and administrative systems, a causal loop diagram is prepared as shown in Fig. 2. The following section discusses the probable solutions to overcoming the path dependencies in complex systems.

3 Conclusion and Recommendations

The path dependencies in the Cauvery River system have prevented the dispute from being resolved for a very long time. To potentially break the feedback loops we need major interventions in our institutions and our thought processes. In a complex system like the Cauvery River basin, we can find "places" and or "points" where a small modification can produce a big change in the outcome [26, 27]. These leverage points are used after the creation of the causal loop diagram described in the previous section. Meadows [26] listed twelve generalized places to intervene in a system. The easiest and first place to intervene in the system, which is also the least effective, is the constants, parameters and numbers. It refers to the tweaking of a system, like turning the faucet for hot/cold water. It refers to the hard numbers like taxes and subsidies which a government can modify to affect change quickly. However, it will be ineffective in breaking the feedback loop in the Cauvery River basin. The most effective leverage point is changing the whole paradigm on which your system is based. This would be most effective but extremely difficult to execute.

The two major path dependencies identified earlier, number 5 and number 2 on [26], are applicable. Number 5 is a system where it will be effective to change the rules of the system (such as incentives, punishments, constraints), and number 2 is the change in the mindset or paradigm out of which the system – its goals, structure, rules, delays, parameters – arises.

Number 2: The non-inclusivity of climate change in any of the policies in the governance structure requires a major change in the goals of the system. Further ideological challenges, like lack of awareness regarding climate change among the general public, and the constant fight among the citizens for basic amenities due to abject poverty in the country need to be addressed. The Cauvery River exerts a heavy influence on the heritage and the culture of the region it flows through, and politicians of both parties have used this against each other to appease its citizens. The major entity affecting the system in Fig. 2 is Climate Change. Including solutions to tackle climate change would help the governments of both provinces to create awareness in the basin. This in turn will weaken the effect it has on the water system. This will increase the probability of the provinces renegotiating the water-sharing rules. The population living in the basin would be aware of the effects of climate change and would be willing to work with the government in finding a solution. This can break the water demand-social unrest-corruption-openness to negotiate feedback loop from Fig. 2.

Number 5: The lack of punishment, in case a province fails to abide by the rules, set by the institutions is a major change required. Furthermore, the constraint of the system regarding technical problems, like hydrological modelling, aquifer division, demarcation, water supply, demand, etc., also fall into this category. The loss of water through inefficiency is unforgivable in water-scarce cities like Bangalore. With an efficient institutional system in place and the punishments it should impose on wastage of resources, the water supply efficiency-revenue generation-corruption feedback loop from Fig. 2 can be weakened. This would work both for Karnataka (domestic water supply efficiency) and Tamil Nadu (irrigation water supply efficiency).

The apparent change will benefit in terms of institutional effectiveness and climate change as can be inferred from the causal loop diagram; however, ideological and cultural challenges are inherent and abstract to the system and can only be experienced [27–30].

The advantages of developing the causal loop diagrams are the identification of the path dependencies and the discussion of which leverage points can be effective. The major disadvantage of only using a causal loop diagram to model a complex system is that it is not possible to encapsulate the system complexities completely. Therefore, this research is the stepping stone to a further system of systems analysis of the system [31].

References

1. Sundararaju, V.: Cauvery: The river that the Tamils thought would never fail. https://www.downtoearth.org.in/blog/water/cauvery-the-river-that-the-tamils-thought-would-never-fail-64973

2. Sharma, A., Schweizer, V., Hipel, K.W.: Cauvery river dispute : a wickedly complex problem. In: Fang, L., Morais, D.C., and Horita, M. (eds.) 20th International Conference on Group Decision and Negotiation, pp. 44.1–44.6. GDN, Toronto, Canada (2020)
3. Sharma, A., Hipel, K.W., Schweizer, V.: Strategic insights into the Cauvery river dispute in India. Sustainability **12**, 1286 (2020). https://doi.org/10.3390/su12041286
4. Pierson, P.: Chapter 1: Positive feedback and path dependence. In: Politics in Time, pp. 17–53. Princeton University Press, Princeton (2004)
5. Hacker, J.S.: Privatizing risk without privatizing the welfare state: the hidden politics of social policy retrenchment in the United States. Am. Polit. Sci. Rev. **98**, 243–260 (2004). https://doi.org/10.1017/S0003055404001121
6. David, P.A.: Path dependence, its critics and the quest for "historical economics." In: Garrouste, P., Ioannides, S. (eds.) Evolution and Path Dependence in Economic Ideas: Past and Present. Elgar Publishing, Cheltenham, England, Cheltenham (2000)
7. Moorthy, N.S.: Delayed monsoon rekindles Cauvery row. http://www.rediff.com/news/2002/jun/25criver.htm
8. United Nations: United Nations Millennium Declaration. New York (2000)
9. United Nations: Millennium Development Goals, UNDP in India. http://www.in.undp.org/content/india/en/home/post-2015/mdgoverview.html
10. UNDP: Sustainable Development Goals (2015)
11. United Nations: THE 17 GOALS, Sustainable Development. https://sdgs.un.org/goals
12. Sachs, J.D., Kroll, C., Lafortune, G., Fuller, G., Woelm, F.: The Decade of Action for the Sustainable Development Goals. Sustainable Development Report 2021, United Kingdom (2021)
13. Government of Karnataka: Karnataka Climate Change Action Plan. , Bangalore, Karnataka, India (2011)
14. Government of Tamil Nadu: Overview, Characteristics, and Status of Water Resources, Chennai, Tamil Nadu, India (2015)
15. Government of Tamil Nadu: Vision 2023 - Strategic Initiatives in Agriculture, Chennai, Tamil Nadu, India (2013)
16. World Population Review: Bangalore Population 2022 (Demographics, Maps, Graphs). https://worldpopulationreview.com/world-cities/bangalore-population
17. Sudhira, H.S., Ramachandra, T.V., Subrahmanya, M.H.B.: Bangalore. Cities. **24**, 379–390 (2007). https://doi.org/10.1016/j.cities.2007.04.003
18. Government of India: The River Boards Act. Ministry of Law, New Delhi, India (1956)
19. Mannheim, K.: Essays Sociology Knowledge. Taylor & Francis, United Kingdom (2013)
20. Hoekstra, A.Y.: Sustainable, efficient, and equitable water use: the three pillars under wise freshwater allocation. WIREs Water **1**, 31–40 (2014). https://doi.org/10.1002/wat2.1000
21. Sears, L., et al.: Jevons' paradox and efficient irrigation technology. Sustainability. **10**, 1–12 (2018). https://doi.org/10.3390/su10051590
22. Sohail, M., Cavill, S.: Water for the poor: Corruption in water supply and sanitation, Leicestershire (2008)
23. The Better India: This Simple Water-Saving Method Has Increased Rice Yield of TN Farmers 10-Fold! The Better India, Chennai, Tamil Nadu, India, pp. 1–2 (2018)
24. Mortimer, C.: Why two Indian states have started rioting over water (2016). https://www.independent.co.uk/news/world/asia/india-riots-water-latest-deaths-wounded-injured-in-bangalore-karnataka-tamil-nadu-a7247226.html
25. First Post: Bengaluru protests have cost Karnataka over Rs 25,000 crore; public transport worst-hit (2016). https://www.firstpost.com/india/bengaluru-protests-have-dented-more-than-just-image-burning-buses-loss-of-business-cost-karnataka-over-25000-crore-3003634.html

26. Meadows, D.: Leverage points: places to intervene in a system (1999). http://drbalcom.pbw orks.com/w/file/fetch/35173014/Leverage_Points.pdf
27. Chan, K.M.A., et al.: Levers and leverage points for pathways to sustainability. People Nat. **2**, 693–717 (2020). https://doi.org/10.1002/PAN3.10124/SUPPINFO
28. Homer-Dixon, T.: Lectures offered during Winter 2019 at the Balsillie School of International Affairs, GGOV622: Complexity and Global Governance (2019)
29. Fischer, J., Riechers, M.: A leverage points perspective on sustainability. People Nat. **1**, 115–120 (2019). https://doi.org/10.1002/PAN3.13/SUPPINFO
30. Abson, D.J., et al.: Leverage points for sustainability transformation. Ambio **46**, 30–39 (2017). https://doi.org/10.1007/S13280-016-0800-Y/TABLES/2
31. Xu, H., Hipel, K.W., Kilgour, D.M., Fang, L.: Conflict Resolution Using the Graph Model: Strategic Interactions in Competition and Cooperation. Springer, Cham (2018). https://doi. org/10.1007/978-3-319-77670-5

Collaborative Decision Making Processes

Visualization of the Social Atmosphere Using Comments on News Sites

Tetsuya Tanaka[1], Madoka Chosokabe[1(✉)] (iD), Keishi Tanimoto[1],
and Satoshi Tsuchiya[2]

[1] Faculty of Engineering, Tottori University, Tottori, Japan
mchoso@tottori-u.ac.jp
[2] Kochi University of Technology, Kochi, Japan

Abstract. When an emergency such as an infectious disease or natural disaster occurs, a negative atmosphere will usually spread throughout society—increasing people's dissatisfaction and anxiety. Because of this, it is rather difficult to thoroughly investigate the actual situation. However, people can post sentimental comments on news sites, allowing for their attitudes either for or against the topics to be better observed. This study extracts the positive, negative, and neutral comments by using sentiment analysis. Then, the social atmosphere is visualized by calculating the approval rating of the comments. This methodology is demonstrated in articles regarding COVID-19. The large volume of comments about two topics, Go To campaigns and PCR tests, were analyzed by using ML-Ask to classify the comments into three categories: negative, positive, and neutral. The results indicate that the social atmosphere about the Go To campaigns tended to be negative.

Keywords: News sites · Text mining · Sentiment analysis

1 Introduction

COVID-19 has been spreading worldwide since December 2019; the pandemic has yet to be fully contained in Japan, as well as in the rest of the world. Under this unprecedented situation, the government needs to take rapid policy measures which can deter the spread of infection and improve people's living quality. In the decision-making process, it is also necessary to reflect on people's opinion on the policy alternatives. However, the situation is changing day by day. How should the government handle its people's needs and then reflect them in its policies?

During the COVID-19 pandemic, people have been forced to follow stay-at-home advisories and maintain the social distancing guidelines; simply put, they have not been able to live their lives as usual. As this situation continues, negative sentiments such as anxiety, frustration, and anger are likely to accumulate within the general populous. In other words, the overall psychological state of the people may be deteriorating. There is a potential danger that these increases in negativity will lead to crowd behavior. In fact, in Japan, panic purchasing of masks had occurred from February to August 2020. How can we prevent this situation from further escalating?

© The Author(s), under exclusive license to Springer Nature Switzerland AG 2022
D. C. Morais and L. Fang (Eds.): GDN 2022, LNBIP 454, pp. 105–114, 2022.
https://doi.org/10.1007/978-3-031-07996-2_8

Before understanding the people's needs, it would be useful to understand their current state of anxiety and dissatisfaction. This study calls the sum of people's emotions, such as anger and joy, within the social atmosphere. So, how do we measure it? A questionnaire survey would be useful in finding results. In an emergency, however, it is difficult to conduct a large-scale survey or collect new data. It is also desirable to incur as little cost (in terms of time and money) as possible.

In recent years, a number of studies have proposed methods for identifying people's opinions from their comments on the Internet, such as Twitter and Facebook. However, this method is limited in that it can only capture the opinions of the people who wrote said comments. However, some news sites provide information on what kind of comments users have posted, as well as how other users have responded to those comments.

By combining this information, we can better understand how people expressed their feelings about a certain social event and how many people agreed with those sentiments. In this study, we develop a method for measuring the social atmosphere from the statements on the Internet and people's responses to them. Specifically, we analyze the sentiment of comments on Yahoo! News and classify them into positive, negative, and neutral comments. In addition, we add people's reactions to the comments—as well as the time—and then clarify which comments have changed and how.

2   Sentiment Analysis

Sentiment analysis is a method that reveals the sentiments and attitudes of the writer of a document by extracting characteristic words or expressions from textual information or speech and then evaluating them comprehensively. It is currently used for reviewing products and services, as well as for analyzing people's reactions to specific events and politics. In this way, sentiment analysis is useful in a variety of situations, including corporate strategy.

There are two policies in sentiment analysis. The first is to divide the text into "positive" or "negative" by giving each word a numerical number and summarizing them in the text. For example, in the dictionary of semantic orientations of words presented by Takamura et al. [1, 2], the most positive word "wonderful" is given a value of 1, and the most negative word "bad" is given a value of -1. Using the dictionary of sentiment, the words in a sentence are assigned numerical values based on the calculated total values of words in the sentence; it then determines whether the sentence is negative or positive.

The second policy is to give the text more sentiments than just two types: negative or positive. For example, there are various types of sentiments; including like, dislike, happy, sad, and so on. Therefore, we extract these sentiments from words in a sentence. The second method is like the first in that it extracts characteristic words from its sentences. It differs from the former in that each word is assigned a category such as "fondness" or "dislike," rather than a numerical value. For example, the word "bad" is categorized as the sentiment "dislike."

When categorizing text, we use the emotive expression dictionary used in ML-Ask. This is a system for sentiment analysis of text in Japanese [3]. It categorizes text to 10 sentiments based on the word or expression in the written text. The 10 sentiments are Gloom, Shame, Anger, Dislike, Fear, Surprise, Fondness, Excitement, Relief, and Joy.

Which words are categorized to which sentiments is defined in the emotive expression dictionary. There is not necessarily one type of emotion that applies to each word or expression. For example, the word "tremble" is assigned to two emotions: "anger" and "fear."

ML-Ask classifies the text into four categories: 1) positive; 2) negative; 3) neutral; and 4) no sentiment. The former 10 sentiments are further classified in 1) to 4), as shown in Table 1. This means that ML-Ask can extract the 10 types of sentiments, as well as divide the negative and positive texts. In this study, we use the emotive expression dictionary in ML-Ask in order to analyze online comments found on news sites.

Table 1. Categories and sentiments by ML-Ask

Categories	Sentiments
1) Positive	Relief, Joy, Fondness
2) Negative	Gloom, Anger, Dislike, Fear
3) Neutral	Excitement, Surprise, Shame
4) None	Not applicable

3   Methodologies

3.1 Outline of Analysis

In this study, we define "social atmosphere" as the attitudes of comments in online news, as well as the approval of the comment by others. To grasp the social atmosphere of COVID-19, using sentiment analysis, we classify online comments of Internet news as positive, negative, or neutral. Then, we add people's reactions to these comments. We will quantitatively reveal the changes in the three types of comments over time.

Yahoo! News has a function that allows people who read the comments to indicate whether they "agree" or "disagree" with the comments made by the users as shown Table 2. In other words, we can see how many people respond to positive (negative, neutral) comments by saying "I agree." From the comments of the extracted emotion groups and people's reactions to them, we try to measure the social atmosphere. Based on the above information, we use the following three indices to visualize the atmosphere as shown Fig. 1.

1. How positive (negative, neutral) the comments are.
2. How many people "approved" the contents of comment.
3. How many people reacted to the comment.

We try to understand the state of the society at hand by calculating the "approval rate" and "sentimental rate of comment" and drawing a bubble chart. The size of the bubble indicates the number of responses; i.e., how many people voted either for or against. For

Table 2. Example of online comments of internet news

Username	Time	Comments	Agree	Disagree
Noname	1/1/22 9:00 AM	I'm sick and tired of the spread of COVID-19	1000	5
Noname	1/3/22 2:00 PM	The government should take action	100	1

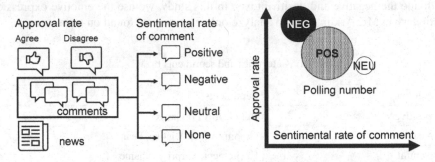

Fig. 1. Overview of the calculation of "approval rate" and "sentimental rate of comment."

example, as shown in the figure on the left in Fig. 2., each bubble has a high approval ratio. This means that there are many supporters for any sentiment; furthermore, we can see that there is a large variety of comments within the society. On the other hand, as shown in the figure on the right in Fig. 2., only one bubble has a high approval ratio. This means that many people support only one emotion and we can see that there is no variety of comments in the society.

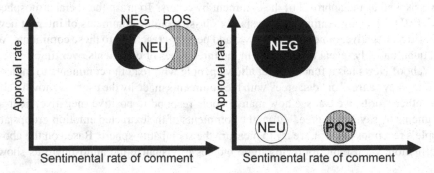

Fig. 2. An example of diversity. In the figure on the left, the three sentiments are in the same place and the size of the circle is the same. On the other hand, in the figure on the right, the three sentiments are in different places and the size of the circle is different.

3.2 Sentimental Rate and Approval Rate in a Comment

First, we calculate the sentimental rate of all comments on news. Let S_{di} denote the sentimental rate of comment i on article d. The sentimental rate $S_{di,j}$ of sentimental category j in comment di among all comments on the news is calculated. Sentimental category j shows positive, negative, or neutral. The sentimental rate $S_{di,j}$ is given as:

$$S_{di,j} = f_{di,j}/W_{di} \tag{1}$$

where $f_{di,j}$ is the frequency of sentimental words v_j in a comment di, and W_{di} is the number of total words in a comment di. There are three types of v_j: positive, negative, and neutral. For example, $v_{positive}$ is defined as including the word, which is categorized by the following sentiments: "Relief," "Joy," and "Fondness." $v_{negative}$ is defined as including the word, which is categorized by the following sentiments: "Gloom," "Anger," "Dislike," and "Fear." $v_{neutral}$ is defined as including the word, which is categorized by the following sentiments: "Excitement," "Surprise," and "Shame." We give one sentimental category to one comment based on the maximum sentimental rate in a comment.

Second, we calculate the approval rate for the comments, which shows how many people agree with the comment. Approval rate A_{di} of comment di is given as:

$$A_{di,agree} = V_{di,agree}/V_{di} \tag{2}$$

where $V_{di,agree}$ is the number of approving votes for comment di and V_{di} is the number of total votes of a comment di. Then, the weighted mean average of sentimental category j is given as:

$$\overline{S_{dj}} = \sum_{i=1}^{N_d} V_{di}S_{di,j} / \sum_{i=1}^{N_d} V_{di} \tag{3}$$

where N_d is the number of total comments on article d. The weighted mean average is of sentimental category j also given as:

$$\overline{A_{dj}} = \sum_{i=1}^{N_d} V_{di}A_{dij} / \sum_{i=1}^{N_d} V_{di} \tag{4}$$

4 Results

4.1 Datasets

Yahoo! News is a website that collects and distributes articles from several newspaper distributors and other sources on the Internet by Yahoo! JAPAN [5]. Yahoo! News articles allow readers to comment on the article; they can then respond with either "agree" or "disagree." We collected Yahoo! News articles and comments that included the keywords "Government" and "COVID-19" in the title or article content. The collection period was set at one year, from June 1, 2020 to May 31, 2021. The collection of comment data was performed by scraping and was implemented in Python.

As a result of the collection, 1,016 articles—with approximately 1.91 million comments—were collected. In this study, we analyzed articles that had more than 10 comments and included "PCR" and "Go To" in the article title—respectively. "Go To" means the Go To travel and eating campaigns, which are a government subsidy for a portion of the cost of travel for individual travelers [4].

First, the comments were analyzed by morphological analysis. Many comments were written on a single article. All comments for each article were analyzed morphologically. Next, we assigned sentiments to the comments. The results of the morphological analysis were compared with the words in the ML-Ask dictionary. If the word was in this dictionary, we assigned a sentiment to it. The sentiment is negative, positive, or neutral. The number of words to be assigned was divided by the total number of words (nouns, adjectives, verbs, adverbs, and inspirations) in order to calculate the degree of emotion. As a result of the calculation, the one with the highest percentage of emotion was chosen as the emotion of the comment.

The sentiment rate of each comment in an article was calculated. The percentage of negative, positive, and neutral for each day were calculated—excluding the comments that had no emotion to be assigned. In addition, we calculated the degree of sentiment and the approval rate for each month. These results were plotted. The sentiment rate on the horizontal axis indicates how many positive (or negative, neutral) words are included in the comment in the case of positive (or negative, neutral) comments. The sentiment rate is a weighted average of the percentage of a certain sentiment in a comment, multiplied by the number of responses to that comment, and divided by the total number of responses to that sentiment in a given month.

4.2 Sentimental Rate of Comment

PCR Test

The number of articles that included "PCR" in the title and received more than 10 comments per day was 32. A total of 9,707 comments were written on 32 articles, of which 2,281 were classified and 7,426 were not classified. The results show that 569 were positive, 1,423 were negative, and 289 were neutral.

Figure 3 shows the sentiment rate about the PCR test and the number of PCR tests per day; it can be seen that the rate of negative (blue line) is large in 2020, while the rate of positive (red line) is large in 2021.

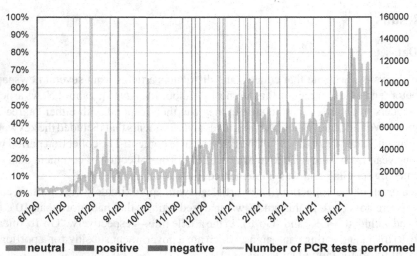

Fig. 3. The sentimental rate about PCR test and the number of PCR tests per day.

Go to Campaign

The number of articles with "Go To" in the title that received more than 10 comments per day was 405. A total of 184,047 comments were written on 405 articles. These comments consisted of 40,002 comments that could be classified and 144,045 comments that could not be classified. The results show that 10,037 were positive, 25,276 were negative, and 4,689 were neutral.

Figure 4 shows the sentiment rate about the Go To campaign and the number of infected persons per day. We can see that there are more articles and a higher percentage of negative comments compared to PCR. Focusing on the percentage of sentiment per day, there were more articles and a higher percentage of negative comments after the Go To campaign was launched in July 2020 and after the Go To campaign was cancelled in some areas in November 2020.

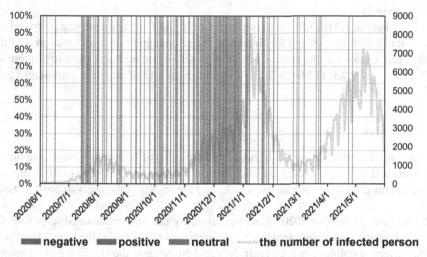

Fig. 4. The sentimental rate about the Go To campaign and the number of infected persons per day

4.3 Change of Social Atmosphere

We focus on the position of the circle and its change in the size of the circle within the bubble chart. The position shows the summary of the positive (or negative, neutral) rate and approval rate of all comments in a month. The size of circle indicates how much the readers voted to the positive (or negative, neutral) comments in a month. The legend symbol, such as "PCR_NEG," means the topics about the article (PCR test or Go To campaign) and the sentiments of comments (negative, positive, and neutral).

PCR Test

The size of the circle of negative comments about the PCR test (black circle) in August 2020 is the largest in a year. This means that many people reacted to the comments in

August 2020, suggesting that people were most interested during this timeframe and assumed a negative reaction.

Focusing on the vertical axis of the bubble chart, the change of PCR circles varies widely than that of the Go To campaign, which means that the attitudes of readers changes—as well. The changes in the horizontal axis of the bubble chart are interesting. The horizontal axis indicates the sentiment rate of comments. The sentiment rate change was greater in both the positive and neutral comments than in the negative comments.

Go To Campaign

Focusing on the change within the size of the circles in the bubble chart, it was shown that the size of the Go To circle is larger than that of the PCR circle. This shows that many people are voting in the comments about the Go To campaign. The result that more people responded to Go To than to PCR suggests that people were more interested in Go To than in PCR.

At the changing in the vertical axis of the bubble chart, the approval rate is around 0.9 without much change—except for September. The change in the horizontal axis of the bubble chart shows that the sentiment rate of positive comments is large in September 2020. From July 2020 to March 2021, the sentiment rate of negative comments (gray circle) has increased.

4.4 Discussion

The results of this study show the trends in the opinions of both those who wrote online comments and those who voted on said comments. The topic on the Go To campaign received many comment votes on the article, and the results of the analysis are discussed below.

In July and November 2020, there were roughly the same number of responses for negative, positive, and neutral sentiments; the approval rate was also the same. There was a wide variety of opinions in society. In July and November, the Go To travel campaign started; in November, the number of infected people increased and there was discussion regarding the implementation of the Go To campaign. These factors are thought to have led to both increased interest and diverse comments from the public.

On the other hand, in August 2020, and January and March 2021, the number of responses to negative comments was high—as is reflected in the size of the circle—and the approval rate was also high. The number of responses to comments on both positive and neutral sentiment was low. This indicates that, at this time, there is a lack of diversity of opinions in society. In January and March 2021, the Go To Travel campaign was temporarily suspended; furthermore, the Go To Eat campaign was suspended in some prefectures. During these months, there was little news about the Go To campaign; it is thought that there was not much interest within society and only the negative comments were supported (Fig. 5).

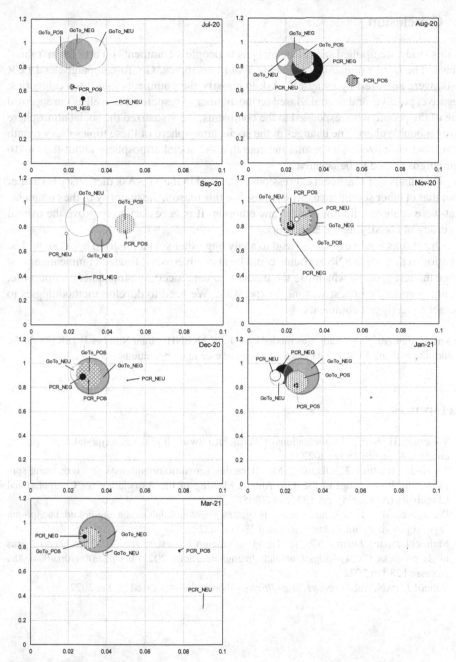

Fig. 5. The changes of "approval rate" and "sentimental rate of comment" by month; the vertical axis is approval rate and the horizontal axis is sentiment rate.

5 Conclusion

In this study, we applied sentiment analysis to people's comments on articles in Yahoo! News. The large volume of comments about two topics, Go To campaigns and PCR tests, were analyzed by using ML-Ask to classify the comments into three categories: negative, positive, and neutral. Based on the number of responses, as well as the approval rate of the people who responded to the comments, we visualized the social atmosphere using a bubble chart. The changes in the social atmosphere of these topics every month were also visualized. The results indicate that the social atmosphere about the Go To campaigns tended to be negative.

The number of sentiment expressions registered in the ML-Ask dictionary is smaller than that of other sentiment analysis tools. For this reason, the majority of the comments that were analyzed did not contain any emotion. It is necessary to improve the overall accuracy of classifying the comments.

Our results could eventually lead to analyzing other social media. Then it should be mentioned that some SNS accounts could be also advertisers, funded commentators, or other interest groups, which use the media to create a certain atmosphere on purpose, ie. not necessarily honest sentiment expression. We need to develop methodologies to extract these rigged comments.

Acknowledgment. This study was funded by JSPS KAKENHI Grant Number JP18K13851. We would like to thank Rentaro Okugawa for his assistance in computations.

References

1. Takamura, H.: Semantic Orientations of Words. http://www.lr.pi.titech.ac.jp/~takamura/pndic_en.html. Accessed 28 Jan 2022
2. Takamura, H., Inui, T., Okumura, M.: Extracting semantic orientations of words using spin model. In: Proceedings of the 43rd Annual Meeting of the Association for Computational Linguistics (ACL2005), pp. 133–140 (2005)
3. Ptaszynski, M.: ML-ask: affect analysis system. http://arakilab.media.eng.hokudai.ac.jp/~ptaszynski/repository/mlask.htm. Accessed 28 Jan 2022
4. Mainichi Japan, Japan's 'Go To Travel' campaign to restart Jan. 2022 or later if virus meds progress (2021). https://mainichi.jp/english/articles/20211111/p2a/00m/0na/004000c. Accessed 28 Jan 2022
5. Yahoo! JAPAN, Yahoo! News. https://news.yahoo.co.jp/. Accessed 28 Jan 2022

On the Dissolution of Three-Party Partnerships via a Buy-Sell Mechanism

Yigal Gerchak(✉) (iD)

Department of Industrial Engineering, Tel-Aviv University, Tel-Aviv, Israel
`yigal@tauex.tau.ac.il`

Abstract. How to dissolve a three-party partnership? Three and more partners partnerships do exist. While the dissolution of two-party partnerships and of analogous business relations was widely discussed, not so for a three-party partnership where the partners initially own arbitrary shares of the firm. Even the possible rules for such dissolution are not quite clear. Assuming some rule, one need to optimize partners' strategies. The efficiency and fairness of such arrangement need then to be investigated. The methods of dissolution that we discuss extend the buy-sell procedure of two partners.

Keywords: Partnership · Dissolution · Three partners · Buy-sell

1 Introduction

Partnership agreements often include a clause specifying how the partnership is to be dissolved if a partner wishes it. We address situations where the business will continue to function while one (or some) of the partners will leave and be compensated, or the share(s) of the partners will change with appropriate payments. If the value of the business is unknown, the partners may have different valuations for it, should they become its sole owners. One method of doing that is a "Knockout Auction", where the partners bid for the (entire) firm, the winner becomes its sole owner, and pays the others their shares of the winning bid [7]. Another is a buy-sell ("shotgun", "Texas shootout", "Bambi") arrangement [9], where one partner (*she*) declares a value for the firm and the other partner(s) (*he*) either sell their shares to her at the declared value, or buy her share of it; if there are three partners, they may also buy each other's share. Indeed while three (and more) partners partnerships exist (e.g., [5]), how to dissolve them and the consequences of the method are not clear.

We propose methods based on the two-partners buy-sell logic for three partners. The methods are a function of prior agreement whether the other two partners are allowed to trade with each other or only with the declarer.

It turns out that the schemes are inefficient. Thus it is possible that the asset will finally be owned by a partner who values it less. However, recent empirical research has shown that bids in two-partners buy-sell experiments [11] are not leading to such undesirable outcome. Second, the two-partners buy-sell procedure provides a systematic advantage to the responder over the declarer and thus is considered to be unfair [14].

With few exceptions, existing literature has analyzed the buy-sell procedure only for the case of two partners, and quite often also assumed that they initially hold equal shares in the asset. This limitation is surprising, since alternative procedures are often discussed in the context of several players, and the similar cake-cutting procedure for fair division has been extended to several players long ago [16]. For a report on real multi-player partnership dissolutions see Wikipedia [17]. Although it cannot be expected that a procedure which sometimes generates inefficient solutions in the case of two partners will lead to efficient outcomes in the case of more than two partners, it is of interest to analyze how the properties of the procedure evolve in the case of more partners. In particular, one could intuitively assume that the advantage of a responder is reduced if the declarer faces several other partners with whom she can trade. Is it so?

In the present article, we therefore study some extensions of this procedure to the case that the asset is initially owned by three individuals. Although the transition from two to three players might seem a miniscule one, adding a third player introduces qualitative changes to the problem which are also relevant for the more general case of n players; like the fact that responses of players to a price stated by the declarer can be heterogeneous. We consider several variants of the procedure, which in turn preserve different properties of the two player case in the more general setting. Given the complexity of the problem, we have chosen a somewhat less general framework than Gerchak and Fuller [9] to model a partner's uncertainty concerning other partners' valuations. We focus mostly on a family of Beta $(\alpha, 1)$, i.e., power (α) distributions with cdf $F(x) = x^{\alpha}, \alpha \geq 1$ on $[0, 1]$ which reduces to a uniform $[0, 1]$ for $\alpha = 1$, and on risk neutral partners.

The paper is structured as follows: In section two, we analyze the two players case in the framework used in this paper to provide a benchmark against which we then compare several procedures for the three player case in section three. Section four concludes the paper by summarizing its main results by providing an outlook onto open research questions.

2   Benchmark: The Two Player Problem

2.1   Notation and General Assumptions

We consider here a situation in which two (or, in the following section, three) individuals $i = 1, 2$ jointly own some asset. The share of partner i in the asset is denoted by s_i, $0 \leq s_i \leq 1$ where $\sum s_i = 1$, and thus, in the two-player case, $s_2 = 1 - s_1$. Each individual holds a private valuation v_i for the entire asset should (s)he become its sole owner. In accordance with most of the literature on similar problems, we assume that these values are independent and identically distributed (i.i.d.) random variables. Furthermore, we assume that this valuation is independent of the actual ownership structure, i.e., shared ownership is assumed to neither create nor destroy value, and each player is able to realize his or her full value of the entire asset once he or she becomes its sole owner. The values v_i are considered to be private information of the individuals. In accordance with most of the literature [1, 6] we further assume that a player's utility function is linear in the asset and money, and that the valuation is expressed in monetary terms.

Each partner has subjective beliefs about the valuations of the other partners. Here we follow the model by Kittsteiner et al. [11] and assume that valuations of all partners are drawn from a power(α) distribution on the zero-one interval, and this distribution is common knowledge of all players. Without loss of generality, we assume that player 1 (P1) is the declarer, and denote the value stated by the declarer by x_1. Since $0 \le v_i \le 1 \forall i$, then $0 \le x_1 \le 1$.

2.2 Model Analysis

P1, who values the entire asset at v_1, initially owns a share s_1 of it. Her initial wealth is therefore $s_1 v_1$. If P2 sells his share at price x_1, player 1 final wealth is $v_1 - x_1(1 - s_1)$, and it owns the entire asset. If P2 decides to buy P1's share, P1's final wealth is $x_1 s_1$. P2 will decide to sell iff $v_2 \le x_1$. We denote the probability that P2 will sell at $p = F(x_1)$, where F is the cdf of P2's valuation of the asset (v_2). In the particular case of the power (α) distribution[1] on $[0, 1], p = x_1^\alpha, \alpha \ge 1$.

The expected profit of P1, if she declares a value x_1, is therefore

$$\pi_1(x_1) = x_1^\alpha \left(v_1 - x_1^\alpha(1 - s_1)\right) + \left(1 - x_1^\alpha\right)x_1^\alpha s_1 - s_1 v_1$$
$$= \left(s_1 - x_1^\alpha\right)\left(x_1^\alpha - v_1\right) - x_1^{2\alpha}. \tag{1}$$

If player 1 states her true value ($x_1 = v_1$), the right side of (1) becomes zero, leading to zero profit for player 1. Player 1 will state a value which is different from v_1 only if this will increase her expected profit above that of stating the true value. This implies that the optimal profit of player 1 cannot be negative.

For a power(α) distribution maximization of $\left(s_1 - x_1^\alpha\right)\left(x_1^\alpha - v_1\right) - x_1^{2\alpha}$ with respect to x_1 shows that a necessary condition for the optimal value to be declared by P1 is

$$-2\alpha x_1^{2\alpha-1} + \alpha(v_1 + s_1)x_1^{\alpha-1} = 0 \tag{2}$$

$$\Rightarrow x_1^\alpha = \frac{v_1 + s_1}{2} \Rightarrow x_1 = \left(\frac{v_1 + s_1}{2}\right)^{\frac{1}{\alpha}} \tag{3}$$

$$\text{so } \pi_1^*(x_1) = \frac{(s_1 - v_1)^2}{4}, \text{ independent of } \alpha \tag{4}$$

P1 thus has an incentive to deviate from her true value v_1. Depending on s_1, she will declare a lower value than v_1 if $s_1 < v_1$, or larger value if $s_1 > v_1$. This conclusion has similarity to a result of McAfee [13] that in the case of arbitrary distribution and equal shares, declarers who have a valuation which exceeds the median of the distribution of valuation will under-declare and declarers whose valuation is lower than the median will over-declare.

An obvious concern when analyzing any mechanism which allocates resources is the efficiency of the resulting allocation. In the present problem, efficiency can be expressed in several ways. Existing literature has focused on the question whether the asset is finally assigned to the player who has the higher valuation. Since we want to compare different settings, we need a cardinal measure of efficiency which goes beyond the binary statement that an allocation is efficient or inefficient.

---

[1] That is actually a beta (α, 1) distribution.

One way of measuring efficiency is to consider the probability of an (in)efficient allocation. To determine this probability, we can represent the problem in v_1/v_2 space as illustrated in Fig. 1. The 45° line separates the regions in which players 1 or 2 have the higher valuation, respectively. The allocation is efficient if in the upper left half of the unit square, the asset is assigned to P2, and in the lower right half to P1. The asset will be bought by P2 if his value is above the purchase price x_1, which for the case of the uniform [0, 1] distribution is represented by the line $v_2 = \frac{v_1 + s_1}{2}$. Therefore, the two shaded triangles represent inefficient allocations. In the left triangle (of diagonal lines), P2 sells his share, although he has a higher value for the asset; in the right triangle P2 buys, although it would be efficient to allocate the asset to P1.

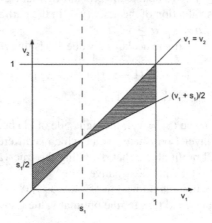

Fig. 1. Efficiency of allocation in the benchmark case

3 The Three Players Problem

We now consider generalizations of the buy/sell procedure to an asset that is initially owned by three partners in shares s_1, s_2 and s_3, where $\sum s_i = 1$. We still assume that one of these partners, P1, takes the role of a declarer, and states a value x_1 for the entire asset. The other two players (2 and 3) are responders, and indicate whether they want to buy or sell shares in the asset at the stated price. If both responders indicate "buy", they split P1's share.

The partners' valuations of the asset, should they become its sole owner are v_1, v_2, and v_3 unknown to the other players.

In the two player problem P2 will trade with P1, and ultimately player one will either own the entire asset or none of it. The same will happen in the three players case, if both responders choose the same alternative (i.e., if both respond "buy" or both respond "sell"). However, if the two responses are different, the situation changes. If the initially set "rules" are that all trades must involve the declarer, the declarer will sell to one responder and buy from the other. At the end, the declarer will thus still hold a fraction of the asset corresponding to the share of the selling responder. This situation can be avoided if trade between responders is allowed. In the case of divergent responses, the responder who wants to sell then trades with the responder who wants to buy, and with the declarer, and the buying responder will thus end up owning the entire asset.

We therefore consider two variants of the procedure in the three player case: Mechanism A, which allows for trade between the two responders, and Mechanism B, in which all trades take place only between the declarer and a responder. We furthermore assume that if both responders indicate "buy", they split the share of the declarer among them. For most of the following analysis, the ratio in which P1's share is split among the responders does not impact on the results.

3.1 Mechanism A

The rules for this procedure are as follows:

1. P1 (the declarer) declares a value.
2. If both responders indicate "sell", P1 receives the shares of both responders at her stated price. If both indicate "buy" they split P1's share.
3. If one of the responders indicates "buy" and the other indicates "sell", the one who indicates "buy" obtains the shares of the two other players for the price stated by the declarer.

This mechanism is very similar to the two-player case. We first show that the responders do not have an incentive to deviate from their valuations, and will indicate "buy" if their value exceeds the stated price and "sell" otherwise. Consider the decision of P2 whether to indicate "buy" or "sell", as shown in Fig. 2.

For P2, the decision of P3 is a random event. In case P2 indicates "sell", he will either sell to P1 or to P3; in any case he receives the value of her share according to the stated price x_1, i.e. $x_1 s_2$.

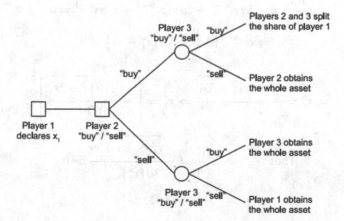

Fig. 2. Decision of P2 in simultaneous model A.

If P2 chooses "buy", and P3 also indicates "buy", he receives some fraction α of the share of P1 at price x_1. His total expected wealth is thus

$$v_2(s_2 + \alpha s_1) - x_1\alpha s_1 = v_2 s_2 + \alpha s_1(v_2 - x_1) \tag{5}$$

If P3 indicates "sell", player 2 obtains the entire asset; her final wealth is therefore

$$v_2 - (s_1 + s_3)x_1 = v_2 s_2 + (s_1 + s_3)(v_2 - x_1) \tag{6}$$

For $v_2 > x_1$, both (5) and (6) are larger than $v_2 s_2$. Otherwise, they are both smaller. Thus, P2 will indicate "buy" iff $v_2 > x_1$ and "sell" otherwise. The same holds for P3. Thus, P1 will sell her share if $x_1 \leq \max(v_2, v_3)$ and buy otherwise. Thus she only has to take into account the distribution of $\max(v_2, v_3)$. This means that the analysis for arbitrary cdf's in Gerchak and Fuller [9] can be directly applied.

For simplicity, assume that the values of both players 2 and 3 are drawn from the same distribution, so for both the probability that they will sell at a given price x_1 is denoted by p. Using a similar analysis as in the two-player case, we can therefore directly write the expected profit, which for a risk neutral player is equivalent to expected final wealth of player 1, as

$$\Rightarrow \pi_1 = \left(s_1 - p^2\right)(p - v_1), \tag{7}$$

which is very similar to Eq. (1) except that first p is replaced by p^2, as P1 will now buy only if both responders are willing to sell (and will sell if at least one wants to buy). Thus the left factor of (7) is larger than of (1). Therefore, for the power (α) distribution, for any $x_1 > v_1^{\frac{1}{\alpha}}$, the expected profit of P1 will be higher in the three player case. This implies that even though the optimal values of x_1 are different, P1's expected wealth *increases* by adding a third player. Specifically,

$$\frac{d\pi_1}{dx_1} = -3\alpha x_1^{3\alpha-1} + 2\alpha v_1 x_1^{2\alpha-1} + s_1\alpha x_1^{\alpha-1} = 0$$

$$\Rightarrow \alpha x_1^{\alpha-1}\left(3x_1^{2\alpha} - 2v_1 x_1^{\alpha} - s_1\right) = 0 \tag{8}$$

$$\Rightarrow x_1^{\alpha} = \frac{v_1 + \sqrt{v_1^2 + 3s_1}}{3} \Rightarrow x_1^* = \left[\frac{v_1 + \sqrt{v_1^2 + 3s_1}}{3}\right]^{\frac{1}{\alpha}}. \tag{9}$$

Thus

$$\frac{dx_1^*}{dv_1} = \frac{\sqrt{v_1^2 + 3s_1} + v_1}{3\alpha\left(x_1^*\right)^{\alpha-1}} > 0. \tag{10}$$

and

$$\frac{dx_1^*}{ds_1} = \frac{1}{2\alpha\left(x_1^*\right)^{\alpha-1}} > 0. \tag{11}$$

So the higher P1 values the asset, and the larger her share, the higher her optimal declaration. The explanation is that if she sells, v_1 has no effect while if she buys her wealth grows in v_1. So, the declaration is increasing in v_1, regardless whether she sells or buys her wealth is increasing in s_1, so does the declaration.

Since we have an explicit solution for x_1, it is possible to calculate the boundary between under and overstating P1's true value in $\frac{v_1}{s_1}$ space:

$$v_1 = x_1 = \left(\frac{v_1 + \sqrt{v_1^\alpha + 3s_1}}{3} \right)^{\frac{1}{\alpha}}, \tag{12}$$

so

$$3v_1^\alpha - v_1 = \sqrt{v_1^\alpha + 3s_1}, \tag{13}$$

that is,

$$s_1 = v_1^{\alpha+1}\left(3v_1^{\alpha-1} - 2\right). \tag{14}$$

Thus $x_1 > v_1$ iff $s_1 > v_1^{\alpha+1}\left(3v_1^{\alpha-1} - 2\right)$.

3.2 Mechanism B

The rules of this procedure are as follows:

1. P1 declares a value x_1.
2. Both responders simultaneously indicate "buyv or "sell".
3. All transactions take place between the declarer and one responder. A responder who indicates "sell" will sell to P1, a responder who indicates "buy" will buy from P1.

If both responders indicate "buy", or both indicate "sell", the outcome is the same as in mechanism A. However, if one indicates "buy" and the other "sell", then at the end the player who has indicated "buy" will own his share and the initial share of P1, while P1 will own the share initially owned by the responder who has indicated "sell".

The responses of players 2 and 3 depend of whether $v_2, v_3 > x_1$ or not. Therefore, we can in general terms write that $p_i = F_i(x_1) = x_1^{\alpha_i}$.

P1 will keep her share only if both other players indicate "sell", and in that case she will end up owning the entire asset. If at least one player indicates "buy", P1 will sell her share (but might obtain another share). We thus obtain the expected wealth of P1 as

$$s_1\left[x_1^{\alpha_2}x_1^{\alpha_3}v_1 + \left(1 - x_1^{\alpha_2}x_1^{\alpha_3}\right)x_1\right] + s_2 x_1^{\alpha_2}\left(v_1 - x_1^{\alpha_2}\right) + s_3 x_1^{\alpha_3}\left(v_1 - x_1^{\alpha_3}\right). \tag{15}$$

The interpretation of Eq. (15) is straightforward:

1. The declarer will retain her share s_1, valued at $s_1 v_1$, only if both responders wish to sell; otherwise, she will obtain the revenue $s_1 x_1$.
2. If P2 is willing to sell (with probability p^{α_2}), then P1 will realize any profit (or loss) from the difference between her valuation v_1 and the price x_1 as applied to the share s_2.

3. The same holds for P3 and his share s_3.

In terms of P1's profit (taking into account that her initial wealth is $s_1 v_1$), we can rearrange (15) in a similar way as Eq. (1). Since we also assume that values of both players 2 and 3 come from the same distribution, we can write $p_2 = p_3 \equiv p$. (i.e., $\alpha_3 = \alpha_2 \equiv \alpha$). So the expected profit of P1 is

$$s_1\left[x_1^{2\alpha} v_1 + \left(1 - x_1^{2\alpha}\right)x_1\right] + s_2 x_1^\alpha (v_1 - x_1) + s_3 x_1^\alpha (v_1 - x_1) - s_1 v_1. \qquad (16)$$

As for any probability $p \geq p^2$, it follows from (1) and (6), that Model B will, under the same condition $\left(x_1 > v_1^{\frac{1}{\alpha}}\right)$, generate a smaller expected profit for P1 than model A.

The expected total profit of P1 is here

$$-s_1 x_1^{2\alpha+1} + s_1 v_1 x_1^{2\alpha} - (1 - s_1)x_1^{\alpha+1} + (1 - s_1)v_1 x_1^\alpha + s_1 x_1 - s_1 v_1, \qquad (17)$$

from which we obtain the first order condition

$$-s_1(2\alpha + 1)x_1^{2\alpha} + 2s_1 v_1 \alpha x_1^{2\alpha-1}$$

$$-(1 - s_1)(\alpha + 1)x_1^\alpha + (1 - s_1)v_1 \alpha x_1^{\alpha-1} + s_1 = 0 \qquad (18)$$

and thus, for $\alpha = 1$

$$3s_1 x_1^2 - 2(s_1 v_1 + s_1 - 1)x_1 + (1 - s_1)v_1 - s_1 = 0, \text{ so } x_1 = \frac{s_1 v_1 + s_1 - 1 \pm \sqrt{\hat{\Delta}}}{3s_1} \qquad (19)$$

$$\hat{\Delta} = s_1^2\left(v_1^2 - v_1 + 4\right) + s_1(v_1 - 2) + 1 > 0 \text{ if } s_1 < \sqrt{\frac{v_1^2 - 4v_1 + 4}{4\left(v_1^2 - v_1\right)}}.$$

$$\text{Thus } x_1 = \frac{s_1 v_1 + s_1 - 1 + \sqrt{}}{3s_1}. \qquad (20)$$

From the first order condition, a truthful declaration requires

$$v_1 = x_1 = \left(\frac{s_1 v_1 + s_1 - 1 + \sqrt{\Delta}}{3s_1}\right)^{\frac{1}{\alpha}}, \quad \Delta = s_1^2\left(v_1^2 + 5v_1 - 2\right) - s_1(5v_1 + 2) + 1 > 0 \text{ if} \qquad (21)$$

$$s_1 < \frac{5v_1 + 2 - \sqrt{3\left(7v_1^2 + 4\right)}}{2\left(v_1^2 + 5v_1 - 2\right)} \text{ or } s_1 > \frac{5v_1 + 2 + \sqrt{3\left(7v_1^2 + 4\right)}}{2\left(v_1^2 + 5v_1 - 2\right)}. \qquad (22)$$

$$v_1^\alpha = \frac{s_1 v_1 + s_1 - 1 \pm \sqrt{}}{3s_1},$$

so,

$$3s_1v_1^\alpha = s_1v_1 + s_1 - 1 \pm \sqrt{}$$

$$\Rightarrow 9s_1v_1^{2\alpha} - 6s_1^2v_1^{\alpha+1} - 6\left(s_1^2 - s_1\right)v_1^\alpha + 3s_1^2v_1 - 3s_1^2 - 3s_1v_1 = 0$$

$$\Rightarrow 3s_1v_1^{2\alpha} - 2s_1^2v_1^{\alpha+1} - 2s_1^2v_1^\alpha + 2s_1v_1^\alpha + s_1^2v_1 - s_1^2 - s_1v_1 = 0 \tag{23}$$

If $\alpha = 1$,

$$3s_1v_1^2 - 2s_1^2v_1^2 - 2s_1^2v_1 + 2s_1v_1 + s_1^2v_1 - s_1 - s_1v_1 = 0$$

$$\Rightarrow \left(-2v_1^2 - v_1 - 1\right)s_1 = -3v_1^2 - v_1 \tag{24}$$

$$\Rightarrow s_1 = \frac{v_1(3v_1 + 1)}{2v_1^2 + v_1 + 1} > 0; \text{ for } s_1 < 1 \text{ need } v_1 < 1. \tag{25}$$

$$s_1 = \frac{v_1\left(3v_1^3 + 2v_1 - 1\right)}{2v_1^3 + 2v_1^2 - v_1 + 1} \tag{26}$$

If $\alpha = 2$,

$$3s_1v_1^4 - 2s_1^2v_1^3 - 2s_1^2v_1^2 + 2s_1v_1^2 + s_1^2v_1 - s_1^2 - s_1v_1 = 0 \tag{27}$$

$$\Rightarrow s_1 = \frac{v_1\left(1 - 2v_1 - 3v_1^3\right)}{-2v_1^3 - 2v_1^2 + v_1 - 1}. \text{ So } 0 \le s_1 \le 1 \text{ iff } v_1 > \frac{1}{2}.$$

P1 might obtain the share of one of the other players and sell her own share, so its total wealth at the end of all transactions will depend also on s_2 and s_3. For simplicity, we assume that $s_2 = s_3 = \frac{1-s_1}{2}$. We have to distinguish among three cases:

1. P1 ends up with the entire asset; that happens with probability p^2, which in case of the power (α) distribution, is $x_1^{2\alpha}$.
2. P1 sells her share, but acquires the share of another player. She thus ends up with a share $\frac{1-s_1}{2}$, which she values at $v_1\left(\frac{1-s_1}{2}\right)$, and some other player, who has an expected valuation of $\frac{1+x_1^\alpha}{2}$, holds a share of $\frac{1+s_1}{2}$. That happens with probability $2x_1^\alpha\left(1 - x_1^\alpha\right)$.
3. Finally, with probability $\left(1 - x_1^\alpha\right)^2$, the two other players end up holding the entire asset, which they value at an expected valuation of $\frac{1+x_1^\alpha}{2}$. Here again it is not necessary to distinguish who owns which share, thus the way in which two buying players split the share of P1 is irrelevant here also.

The expected total wealth for P1 is

$$E[V] = x_1^{2\alpha}v_1 + 2x_1^\alpha\left(1 - x_1^\alpha\right)\left(\frac{1-s_1}{2}v_1 + \frac{1+s_1}{2} \cdot \frac{1+x_1^\alpha}{2}\right) + \left(1 - x_1^\alpha\right)^2\frac{1+x_1^\alpha}{2} \tag{28}$$

Function (28) is increasing for $x_1^\alpha > 2(1-s_1)v_1 + \frac{1}{2}(1-3s_1)$, (i.e., for $x_1 > [2(1-s_1)v_1 + \frac{1}{2}(1-3s_1)]^{\frac{1}{\alpha}})$, and decreasing otherwise. It is convex for $x_1^\alpha > \frac{2(1-s_1)v_1+\frac{1}{2}(1-3s_1)}{s_1+1}$, and concave otherwise. Thus its maximum is either at $x_1^\alpha = \frac{2s_1v_1-1-\sqrt{\Delta}}{3s_1}$ or at $x_1 = 1$. Here $\Delta = 4s_1^2v_1^2 + 2s_1(1-3s_1)v_1 + 3s_1^2 + 1(> 0)$.

3.3 Comparison of Mechanisms

We can now compare the two three-player scenarios to the two-player one in terms of the values being declared and their deviations from player 1's true valuation, as well as in terms of efficiency and fairness. For this analysis, we will mainly take the perspective of player 1, who knows her s_1 and v_1. We will therefore analyze these questions by determining regions in $\frac{s_1}{v_1}$ space which exhibit different properties.

By equating the declared value x_1 from Eqs. (1), (7) and (20) to v_1, it is possible here to determine for all three models the combinations of v_1 and s_1 for which the declarer specifies her true value, and thus to separate the $\frac{s_1}{v_1}$ space into regions of over- and under-declaring.

Clearly, the introduction of a third player increases the incentive for over-declaring relative to the true value and reduces the incentive for under-declaring. This effect is present in model B, and is even more pronounced in model A. For it we check when

$$\frac{v_1 + \sqrt{v_1^2 + 3s_1}}{3} \geq \frac{v_1 + s_1}{2} \tag{29}$$

i.e.,
$$2\sqrt{v_1^2 + 3s_1} \geq v_1 + 3s_1 \tag{30}$$

i.e.,
$$4\left(v_1^2 + 3s_1\right) \geq v_1^2 + 6s_1v_1 + 9s_1^2 \tag{31}$$

i.e.,
$$v_1^2 - 2s_1v_1 + s_1(4 - 3s_1) \geq 0, \tag{32}$$

$$\Rightarrow v_1 = s_1 \pm 2\sqrt{s_1(s_1 - 4)} \geq 0 \text{ for all } s_1 \text{ and } v_1. \tag{33}$$

The same conclusion can be drawn for mechanism B.

It is intuitively plausible that the presence of a third player increases the incentive for over-declaring. The declarer can obtain a positive profit either by selling her share at a price which is higher than her true value (which is possible if she over-declare or by buying the others' shares at a price which is lower than her value (which requires under-declaring). In model A, the behavior of the declarer is mainly driven by the distribution of the maximum of the valuations of the two other players. Thus, selling and over-declaring is the more likely scenario in this model, and which creates an incentive for over-declaring also in model B. The declarer might still have to buy the share of a responder who wants to sell (even if the other responder buys the declarer's share), which weakens the incentive for over-declaring compared to model A.

This additional over-declaring is also profitable for P1. In case of two players, those players who tend to over-declare (i.e., players who hold a comparatively large share in the asset, but have a low valuation) tend to profit most from adding a third player. On the other hand, players who have a small share and a high valuation (and who consequently under-declare) suffer a loss (in expected value) from adding a third player.

4 Conclusions

Extending the buy-sell procedure to the case of three players is not straightforward. One of the main goals of the buy-sell procedure is to allow one partner, who for some reason might not be satisfied with the current state of affairs, to either completely terminate her membership in the partnership, or to obtain full ownership of the previously shared asset. As we have seen, this requirement is not compatible with the basic rule of buy-sell that all trades take place between the declarer and the other partners. We therefore have analyzed two variants of the three-player procedure, which either guarantee that the declarer ultimately owns all or nothing, or alternatively allow only trades between the declarer and other partners.

An intuitive assessment of the differences between a two- and three-player problems might lead to the conclusion that with three players the declarer must be in a better position than in the two-player case because she could in some way exploit competition between the other two players. Our analysis has indicated that this might not be the case. To the contrary, the presence of a third player might enable a responder who is in a weak position (because he has a rather low valuation of the asset), to "free ride" on the higher valuation of the other responder and sell his share in the asset for a higher price. If one takes into account that the expected highest valuation of the asset is higher in the case of three players than it is in the case of two players, there is also no gain in terms of efficiency.

Therefore, we conclude that within the framework of our analysis, the disadvantages of the buy-sell procedure in the case of two players remain, and to a certain extent are even aggravated, if a third player is added. Still, the buy-sell procedure is an important clause in many actual partnership contracts, so the question how these problems could be overcome remains an important one. It remains to explore the nature of actual buy-sell clauses in three-partner partnerships.

It will also be of interest to compare the magnitudes of x_1^* and the resulting expected profits of P1 for mechanisms A and B.

Our analysis has considered two specific variants of the buy-sell procedure for three players. Both variants were characterized by the fact that only one player takes the role of a declarer, and that both responders reply simultaneously to the declarer's statement. Clearly, many other variants of the procedure are possible. In particular, players could act sequentially, either in the form that a declarer interacts with the two responders one after the other, or in the form of repeated two-player interactions in which one out of two players first obtains the share of the other, and then deals with the third partner. Such sequential procedures would further complicate the analysis, since each transaction reveals some information about the valuations of the transacting partners, which could be used strategically by partners stepping in at a later stage.

Another modification could involve the specification of two values, one at which the declared is willing to buy, another one at which she is willing to sell. Such an approach would obviously remove the impression that the declared value should reflect the declarer's true value, and thus probably opens more room for strategic behavior by the declarer. This could offset the disadvantage of the declarer in the standard model.

Our analysis has provided a framework for, and a first step in the analysis of, multi-partner buy-sell procedures, in which other models could be built to provide a better understanding of such procedures and their consequences.

Acknowledgements. I wish to thank Rudolf Vetchera for his extensive help in this research.

References

1. Athanassoglou, S., Brams, S.J., Sethuraman, J.: A note on the inefficiency of bidding over the price of a share. Math. Soc. Sci. **60**(3), 191–195 (2020)
2. Brams, S.J., King, D.L.: Efficient fair division: Help the worst off or avoid envy? Ration. Soc. **17**, 387–421 (2005)
3. Brooks, R.R., Landeo, C.M., Spier, K.E.: Trigger happy or gun shy? Dissolving common-value partnerships with Texas shootouts. Rand J. Econ. **41**(4), 649–673 (2010)
4. Cramton, P., Gibbons, R., Klemperer, P.: Dissolving a partnership efficiently. Econometrica **55**(3), 615–632 (1987)
5. Canadian Press, February 13. Stunning deal sees Leafs buy Raptors. (1998)
6. de Frutos, M.A., Kittsteiner, T.: Efficient partnership dissolution under buy-sell clauses. Rand J. Econ. **39**(1), 184–198 (2008)
7. Engelbrecht-Wiggans, R.: Auctions with price-proportional benefits to bidders. Games Econ. Behav. **6**(3), 339–346 (1994)
8. Gerchak, Y.: Decision analytic approach to knockout auctions. Decis. Anal. **5**(1), 19–21 (2008)
9. Gerchak, Y., Fuller, J.D.: Optimal value declaration in "Buy-Sell" situations. Manage. Sci. **38**(1), 48–56 (1992)
10. Gibbons, R.: Game Theory for Applied Economists. Princeton University Press, Princeton (1992)
11. Kittsteiner, T., Ockenfels, A., Thral, N.: Partnership dissolution mechanisms in the laboratory. Econ. Lett. **117**(2), 394–396 (2012)
12. Konow, J.: Which is the fairest one of all? A positive analysis of justice theories. J. Econ. Literature **41**, 1188–1239 (2003)
13. McAfee, R.P.: Amicable divorce: dissolving a partnership with simple mechanisms. J. Econ. Theor. **56**(2), 266–293 (1992)
14. Morgan, J.: Dissolving a partnership (un)fairly. Econ. Theor. **23**(4), 909–923 (2004)
15. Raith, M.G.: Fair negotiation procedures. Math. Soc. Sci. **39**(3), 303–322 (2000)
16. Steinhaus, H.: Sur la division pragmatique. Econometrica **17**, 315–319 (1949)
17. Wikipedia org/wiki/Maple_Leaf_Sports_%26_Entertainment
18. Zukerman, M., Mammadov, M., Tan, L., Ouveysi, I., Andrew, L.L.: To be fair or efficient or a bit of both. Comput. Oper. Res. **35**, 3787–3806 (2008)

Author Index

Printed in the United States
by Baker & Taylor Publisher Services

Printed in the United States
by Baker & Taylor Publisher Services